D1135373

MAPPING THE DETERMINANTS OF
SPATIAL DATA SHARING

For Carlota

Mapping the Determinants of Spatial Data Sharing

UTA WEHN DE MONTALVO
TNO-STB, The Netherlands

ASHGATE

Published by
Ashgate Publishing Limited
Gower House
Croft Road
Aldershot
Hants GU11 3HR
England

Ashgate Publishing Company
Suite 420
101 Cherry Street
Burlington, VT 05401-4405
USA

Ashgate website: http://www.ashgate.com

British Library Cataloguing in Publication Data
Montalvo, Uta Wehn de
        Mapping the determinants of spatial data sharing
        1. Geographic information systems
        I. Title
        910.2'85

Library of Congress Control Number: 2003101438

ISBN 0 7546 3475 2

Printed and bound in Great Britain by MPG Books Ltd, Bodmin, Cornwall

Contents

List of Figures

List of Tables

Preface

Whilst working on 'Knowledge Societies' with its focus on ICTs and developing countries,[1] I became interested in Geographic Information Systems (GIS) and their potential application to a whole range of development activities. Initially, the endeavour received mixed encouragement. An established geographer at the university commented that 'People in developing countries don't use GIS, they just draw maps in the sand'. As it turned out, his judgment was rather wrong. GIS were, and are, being used in many developing countries and examples of sophisticated use, by locals and development organisations alike, abound. As I continued to pursue my interest, I realised that one of the main issues was related neither to the software, hardware nor the capabilities required to run GI systems, but to bottlenecks in the access to, and the availability of, the data required to get meaningful output from a GIS.

Spatial data infrastructures to facilitate collaboration and distribution of spatial data through sharing were becoming an increasingly 'hot topic' in the GIS scene in many countries, both developing and industrialised. Yet most puzzling to me was how it could just be assumed that people actually *want* to co-operate and share data? Is the existence of a 'sharing culture' not just being taken for granted? Explaining the research to anyone not even remotely concerned with, or interested in, GIS, typically yielded the response: 'Of course nobody wants to share!', implying 'Why are you even bothering?' Although the question of whether or not people are willing to share is interesting in itself, the reasons behind *why* they do or do not want to share are of crucial importance. A better understanding of the motivations to share and, perhaps even more importantly, the motivations *not* to share, can help one to relate better to important stakeholders inside and outside the GIS world. This means that it is possible to go beyond mere articulation of the benefits and to actually address the obstacles to data sharing. Rather than preaching to the converted, these insights can help achieve recognition and acceptance of the importance of data sharing beyond the GIS scene to accomplish the effective implementation of spatial data infrastructures.

The data sharing issue that is the topic of this book is a matter of interest for developing and industrialised countries alike. As such, the book primarily addresses those involved in the implementation of spatial data infrastructures at

---

[1] Mansell, R. and Wehn, U. (eds) (1998) *Knowledge Societies – Information Technology for Sustainable Development*, prepared under the guidance of the United Nations Commission of Science and Technology for Development, Oxford: Oxford University Press.

local, national and regional levels, whether they are part of more formal, already established directorates or are informal groups of interested parties just setting out. Owing to the pervasiveness of the data sharing issue and the multi-disciplinary approach adopted herein, it should also be of interest to academics, researchers and students in areas of innovation studies, social psychology and economics.

A book about sharing and co-operation is, of course, developed through co-operation and support. I wish to thank all those interviewed in the various organisations in South Africa who gave so much of their time and provided input for this research. One of the most challenging aspects of the empirical research was the length of the questionnaire (which ran to 25 pages!). By putting aside their work for an hour (or more) to answer the questions, the respondents demonstrated that research requiring this level of detail can be done.

I gratefully acknowledge financial support for this research from the Economic and Social Research Council in the UK and for the preparation of the camera-ready-copy by TNO-STB. I am also grateful to the EDIS conference committee who allowed me to distribute questionnaires at the EDIS conference in July 1999 in Pretoria.

I owe particular thanks to Dr Bob Day who, at the time, was at Mikomtek, Council of Scientific and Industrial Research, and to his secretary, Olga Crawford, for providing logistical support during the fieldwork in South Africa. I also thank Professor Robin Mansell for her commitment to this research and her abundant and constructive comments. I am grateful to Professor Nick von Tunzelmann for taking the time to discuss my research on many occasions. I owe my appreciation to my husband, Carlos, who has been an invaluable source of intellectual and emotional support. Help with editing and the technical aspects of bringing the book to print were provided respectively by Cynthia Little and Holger Lütters and is much appreciated.

Dr. Uta Wehn de Montalvo
Delft, May 2003

List of Abbreviations

Σ	Sum
Σaltern. sources1	Scale to assess perceived scope of alternative sources of spatial data for sharing partners
Σaltern. sources2	Scale to assess perceived scope of alternative sources of spatial data for the organisation
Σbenefits	Scale to assess perceived savings implications
Σcontrol aspects	Scale to assess perceived control over internal spatial data
Σcosts	Scale to assess perceived costs implications
Σdependence aspects	Scale to assess perceived dependence on external spatial data
ΣGIS community	Scale to assess perceived GIS community pressure
Σimportance of external data	Scale to assess perceived importance of external spatial data
Σimportance of internal data	Scale to assess perceived importance of internal spatial data to sharing partners
Σinstitutional	Scale to assess perceived institutional pressure
Σinter-org relations	Scale to assess perceived outcomes for inter-organisational relationships
Σit skills	Scale to assess perceived IT skills
Σknowledge creation	Scale to assess perceived knowledge creation outcomes
Σmarket	Scale to assess perceived market pressure
Σmetadata	Scale to assess perceived metadata-related skills
Σmoral norms	Scale to assess perceived moral norm pressure
Σnegotiation	Scale to assess perceived negotiation skills
Σnetworking	Scale to assess perceived networking and teamworking skills
Σopportunities	Scale to assess perceived opportunities for spatial data sharing
Σorganisational activities	Scale to assess perceived outcomes for organisational activities

Σorganisational pressure	Scale to assess perceived organisational pressure
Σpast experience	Scale to assess past experience with spatial data sharing
Σresource control	Scale to assess perceived control of resources for sharing
Σself sufficiency	Scale to assess perceived self-sufficiency and position in terms of spatial data
Σsharing partners	Scale to assess perceived dependence on sharing partners
Σsocial outcomes	Scale to assess perceived social outcomes
Σspatial data outcomes	Scale to assess perceived spatial data-related outcomes
Σspatial data skills	Scale to assess perceived spatial data-related skills
A	Attitude
AGI	Association for Geographic Information
be	behavioural belief combined with outcome evaluations
CIE	Center for International Economics
cp	control belief combined with perceived control
CSIR	Council for Scientific and Industrial Research (South Africa)
EDIS	Earth Data Information Systems Conference
ESRI	Environmental Systems Research Institute
FGDC	Federal Geographic Data Committee (US)
GIS	Geographic Information System
GSDI	Global Spatial Data Infrastructure
INEGI	National Statistics, Geography and Informatics Institute (Mexico)
INSPIRE	INfrastructure for SPatial InfoRmation in Europe
MSC	Mapping Science Committee (US)
nm	normative belief combined with motivation to comply
NGO	non-governmental (not-for-profit) organisation
NIST	National Institute of Standards and Technology (US)
NSDI	National Spatial Data Infrastructure
NSIF	National Spatial Information Framework (South Africa)
OGC	Open GIS Consortium
p	Level of significance

PBC	Perceived (behavioural) control
PSP	Social pressure
r	Index of bivariate correlation (Pearson)
SD	Spatial Data
SDI	Spatial Data Infrastructure
TPB	Theory of Planned Behaviour
TRA	Theory of Reasoned Action
W	Willingness to engage in spatial data sharing across organisational boundaries

Chapter 1

Introduction

With the increasing use of Geographic Information Systems (GIS) in industrialised and developing countries, the availability of spatial data has become an issue that affects many organisations. Spatial data initiatives are reliant on the willingness of different organisations to engage in spatial data sharing in order to be effective in overcoming bottlenecks in the availability of spatial data. Yet the determinants of spatial data sharing behaviour by different types of organisations have not been dealt with in a systematic fashion and the existing literature on spatial data sharing offers only anecdotal insights into such behaviour. The lack of knowledge about what determines the willingness to share spatial data makes it difficult to address the possible bottlenecks effectively. The research presented in this book employs a social psychology approach to systematise the determinants of organisations' spatial data sharing behaviour and to address the central research question: What appear to be the factors that influence the willingness or resistance of key individuals within organisations to engage in spatial data sharing across organisational boundaries? The book also examines the actual spatial data sharing behaviour of organisations. An improved understanding of the importance of various factors can be used to motivate greater spatial data sharing among organisations at local, regional, national or global levels.

In South Africa, as in many other countries, the National Spatial Information Framework (NSIF) is dominated by the public sector with some private sector involvement. The aim of such frameworks is to reduce high data duplication costs and data availability bottlenecks. By focusing the NSIF predominantly on costs, procedures, clearinghouses and technical standards for spatial data and metadata, current policy developments may be failing to take into account the need to foster a spatial data sharing 'culture'. Such a culture may be expected to increase the likelihood that individuals in different organisations will have a greater propensity to engage in spatial data sharing. The principal focus of this research is an investigation of the nature of the conditions under which the different actors involved with spatial data are likely to be willing to share the data across organisational boundaries. These results are expected to encourage policy developments that promote a behaviour that has the potential to increase the use of an information and communication technology application that is widely regarded as being a useful tool for decision-makers in developing countries (Schwabe *et al.*, 1998).

Background

This study is located in the growing body of research into the socio-economic and political transformations that are accompanying the diffusion of advanced information and communication technologies (ICTs) in developing countries. A specific example of these technologies is Geographic Information Systems (GIS). These are computer systems that can combine spatial and socio-economic data from numerous different data sets. There is considerable evidence that suggests that when GIS are used effectively they can support the activities of monitoring, planning, and policy design in areas of natural and other resource management and for the development of various kinds of physical infrastructures such as transport, utilities and telecommunications.

Spatial data are expensive resources and, particularly in developing countries, appropriate spatial data may not always be readily available.

> As the technology of GIS becomes ever more ubiquitous and apparently ever easier to use, other factors condition its development. Perhaps the key factor is the availability of 'content' - the data and information which act as 'fuel' for geographical information and other computer systems. The existence of such data, their currency, accuracy, and consistency, their availability and price and the terms and conditions of their use, are now all major factors in determining the utility, cost, and effectiveness of GIS. (Rhind, 1999: 767)

Some of the possible bottlenecks in the availability of, and access to, spatial data have been found to arise from insufficient flows of information and poor data management (Onsrud and Rushton, 1992).

> Since acquisition of certain types of geographical information or geospatial data is difficult and/or costly to collect for many individuals, this concern has translated into how to get the data from other parties - where to find them; how to obtain their costs, currency and reliability; and the terms under which they can be used and the liabilities incurred. (Rhind, 1999: 767)

Potential spatial data users may have difficulty finding, or gaining access to, relevant spatial data because of a lack of institutional co-ordination. In many instances, potential spatial data suppliers may not be aware of what geographic information they have accumulated or who might need it. Poor management of data resources hampers dissemination of spatial data (Campbell, 1991). It results in duplication, inefficiency and lost opportunities for the use of spatial data to assist in solving problems. The relevant spatial data may need to be recaptured at great expense or projects involving the use of such data may have to be abandoned.

As the importance of using spatial data to address complex social, environmental and economic issues continues to grow (Nebert, 2000; Longley *et al.*, 1999), the need to access, integrate and use spatial data from disparate sources to guide decision making is being recognised by public and private sector organisations.

National, regional, and even global efforts are underway to better co-ordinate the collection, distribution and use of digital geographic information. These initiatives are actively promoting access to, and re-use of, spatial data by encouraging collective conventions and agreements.

Only through common conventions and technical agreements will it be easily possible for local communities, nations and regional decision-makers to discover, acquire, exploit and share geographic information vital to the decision process. (Nebert, 2000: 7)

The underlying principle that is being advocated by participants in these initiatives - spatial data sharing - involves exchange relationships that may or may not include financial payment or payment in kind and that entail making spatial data accessible to, or from, other parties under certain terms and conditions. However, there are major questions about the willingness of the individuals involved to engage in spatial data sharing. The new common conventions and agreements address issues of spatial data sharing only at the technical level. However, the success of these initiatives is likely to depend significantly on the *willingness* of key individuals within different organisations and institutions involved with spatial data to co-operate and to engage in spatial data sharing activities. In the light of experience so far suggesting that there are considerable non-technical barriers that are likely to discourage data sharing, it is important to systematically explore and assess what the factors are that contribute to such resistance. To this writer's knowledge, there have been no systematic investigations of the non-technical issues - i.e. the behavioural and perceptual issues - that are likely to strongly influence data sharing behaviour.

For example, organisations making more spatial data accessible than they receive may perceive that they are losing out while the data receivers may believe they are gaining from an exchange relationship. Alternatively, it may be believed that the distribution of spatial data to decision makers provides an opportunity to gain influence with those decision makers, thus reducing resistance to data sharing. These are only a few of the complex rationales and beliefs that individuals within organisations may hold about the advantages and disadvantages of sharing spatial data.

In order to investigate whether improvements in spatial data sharing are likely to follow from national initiatives that are designed to promote the technical standard setting that must be in place to facilitate spatial data sharing, the establishment of the necessary clearinghouses, and consistent pricing policies for national spatial data infrastructures, it is argued in this book that it is *also* necessary to identify the existing incentives and disincentives for participating in spatial data sharing that may be present within communities of actors and the nature of the interactions of these incentives and disincentives involving the generation and use of spatial data. The results of this research are intended to provide an improved basis for generating specific guidelines as to how the social aspects of spatial data sharing can be addressed more effectively within the framework of these initiatives.

Theoretical Location of the Research

This study is intended to extend research in the field of studies concerned with the diffusion of technological innovations (Rogers, 1968, 1995; Rosenberg, 1976; Kline and Rosenberg, 1986; Coombs *et al.*, 1987; Metcalfe, 1981, 1988; Antonelli, 1995; Sarkar, 1998). The research is designed to consider the social preconditions for the diffusion of a specific technology. These preconditions are likely to increase the possibility that users of the technology will behave in ways that are consistent with maximising the potential benefits of the application of the technological innovation. The specific technology considered in this study is GIS. Spatial data are indispensable for the use of GIS and they are increasingly used by public and private sector organisations.

The growing importance of spatial data availability and data access for the effective use of GIS has begun to generate a branch of literature that focuses directly on the issue of spatial data sharing (Onsrud and Rushton, 1992, 1995; Rhind, 1992, 1997, 1998, 1999; Obermeyer and Pinto, 1994; Obermeyer, 1995; Pinto and Onsrud, 1995; Burrough and Masser, 1998; Masser 1999; Nedovic-Budic and Pinto, 1999a,b, 2000, 2001). This work provides a basis for developing a definition of spatial data sharing for the purpose of this study. While this work offers insights into some of the factors that may influence individuals' engagement in spatial data sharing activities, it does not provide a systematic account of the social aspects of this behaviour or the beliefs that individuals involved in the generation and use of spatial data may hold. Therefore, the conceptual framework for this study draws on the Theory of Planned Behaviour (TPB) developed by Ajzen (1991) in the field of social psychology. This theory is adapted and applied to provide a framework for examining the factors that influence individuals' willingness to engage in spatial data sharing. The main theoretical constructs that are developed and operationalised in the study are the elements of 'attitude', 'subjective norm' and 'perceived behavioural control'. Evidence with respect to these constructs is expected to indicate how they influence the willingness of individuals (or organisations) to engage in a particular behaviour. According to the TPB, beliefs are assumed to provide the basis for the formation of 'attitude', 'subjective norm' and 'perceived behavioural control'.[2]

Using the TPB as a conceptual framework, a model is developed and then applied in a real-life situation to capture perceptions of the willingness of

---

[2] The TPB is one of a class of theories specifically intended to explain the relationships between beliefs and the formation of attitudes. There are those for whom the expression of beliefs by individuals may be deemed to be only an indirect, or even misleading, indication of attitudes, and who would argue that the whole social context influences what individuals will claim about their beliefs at different times and for various purposes, i.e. the expressed beliefs may be a reflection of political motivations rather than the actual beliefs (see, for example, a discussion in Sudman and Bradburn (1982)). As far as possible, the research design and methodology for this research was developed to provide a large enough sample such that controls for inconsistencies would be included. For the purpose of this particular study, debates about the underlying motivations for the expression of beliefs are excluded from the analysis.

individuals embedded in organisations to engage in spatial data sharing. In the construction of the model, the antecedents of the main components of the model are traced to underlying beliefs and a variety of bodies of literature are drawn upon to operationalise beliefs where insights gained from personal interviews were not sufficient.

For example, actors who are asked about their engagement in spatial data sharing may anticipate that if they share the data they may experience a loss of control over the data their organisation possesses. In order to translate this perception into a rating of a belief, the study incorporates insights from resource dependence theory (Pfeffer and Salancik, 1978; Alter and Hage, 1993; Mizruche and Galaskiewicz, 1993). Resource dependence theory focuses on the extent of dependence that organisations experience by examining the actual amount of control other parties have over access to a required resource. The evaluation of the extent of control over resources can be conceptualised so as to provide means of assessing the extent to which actors perceive they may experience a *loss of control* over their spatial data and how this, in turn, influences their attitude towards sharing spatial data. The resource in question, in this case, is the spatial data that are available to be exchanged and the focus of the evaluation is on the degree of discretion that key individuals perceive their organisations have over the allocation and use of the spatial data possessed by their organisation.

Another instance of the use of ideas drawn from the resource dependence approach is in the consideration of the dependence of an organisation on, and its control over, spatial data. The resource dependence approach posits a link between dependence on other organisations and an organisation's behaviour (Pfeffer and Salancik, 1978). It has also been argued that organisational activities are not entirely determined by the environment within which the organisation operates and, therefore, that an organisation's relative self-sufficiency affects the extent of its dependence on the environment (Alter and Hage, 1993). Furthermore, the resource dependence approach focuses on the role of resources in inter-organisational relationships and provides insights into practical means of assessing the dependence for resources of one organisation on another. Concepts developed in this literature concerning both external and internal factors that influence organisational behaviour are used in this study as a basis for measuring the dependence of organisations, as perceived by key individuals, on external sources of spatial data, and their perceived control over internal spatial data resources.

Intangible consequences of spatial data sharing may also influence the willingness of key individuals within organisations to engage in spatial data sharing. Intangible effects of sharing spatial data, such as the opportunity to gain new insight or the fear of diluting ideas, are considered within the context of the factors at play in the process of knowledge creation. Various modes of knowledge creation have been proposed (Nonaka, 1994) and these are utilised as a means of determining the extent to which aspects of the knowledge creation process in the context of spatial data sharing across organisational boundaries may be perceived as threats or opportunities for the individuals involved. Finally, research on spatial data sharing provides a basis for operationalising the beliefs of individuals that are identified from interviews with actors involved in the GIS community.

Research Questions

The principal aim of this research is to examine the nature of the conditions under which the different actors involved with spatial data are likely to be willing to share the data across organisational boundaries to overcome bottlenecks in the availability of, and access to, spatial data so that GIS can be used more effectively.

To this end, the central research question addressed in this study is: What appear to be the factors that influence the willingness or resistance of key individuals within organisations to engage in spatial data sharing across organisational boundaries? The research also examines the actual spatial data sharing behaviour of the organisations included in the study.

The results of the research offer a systematic basis for deepening understanding of how spatial data sharing can be fostered so that bottlenecks in the availability of, and access to, such data may be overcome through policy initiatives. The investigation provides insights into why key individuals within GIS-using organisations believe their organisations are likely to gain or lose from engaging in spatial data sharing; whether they think those who are important referents for their organisations want them to share spatial data; and what control in terms of skills, capabilities, and resources they perceive that their organisations have over the sharing of spatial data. When a profile of such perceptions is established, it provides a basis for addressing the issue of spatial data sharing more effectively at the policy making level.

Research Design

This research extends and applies the TPB as a conceptual framework to provide a systematic account of the factors that appear to influence the willingness of individuals (and organisations) to share spatial data. This theory was originally developed to understand individual behaviour but it can also be applied to organisations by focusing on key individuals within organisations (Elliot *et al.*, 1995; Staw, 1991). The selection of key individuals from each of the organisations in the sample constructed for this study is based on the assumption that these decision makers can be expected to have a reasonably accurate understanding of their organisation's position on spatial data sharing matters. Therefore, their responses to questions about their beliefs and behaviour provide a good indicator of their organisations' behaviour. In this respect, the individual's perception can be taken to be consistent with the collective view of an organisation.

The representation by individuals of their views and the interpretation of these views as indicators of likely organisational behaviour provides the basis for the research strategy adopted here. By studying key decision makers embedded in their organisations rather than focusing in depth on all aspects of individual behaviour within an organisation (Staw, 1991), the research is not limited to investigating a very few organisations and there is a basis for generalisations to be drawn from the research results that is stronger than would be the case if the research strategy had employed a small number of in-depth case studies. A range of individuals and organisations has been included in the sample for this study with the goal of examining their dispositions towards spatial data sharing across organisational boundaries. This research strategy provides a means to extend the analysis to encompass a larger arena (Markus and Robey, 1988), in this case, to encompass the full scope of a single national spatial data infrastructure and a representative sample of its participants. However, although one goal of the study was to extend the scope of the analysis, there was also a need to balance this goal with that of achieving a reasonable assessment of the complexity of the social dimension of the behaviour in question. The need to conduct the research in sufficient depth to reveal substantial detail about the beliefs and perceptions of a representative sample of actors and organisations meant that the empirical research focused on one country, South Africa, rather than on several countries which would have provided a basis for inter-country comparison.

The research methodology was implemented in two stages of empirical research. The qualitative stage involved semi-structured interviews, the results of which were analysed and complemented by insights drawn from the appropriate theoretical and empirical literatures to build a conceptual model of the willingness of individuals within organisations to share spatial data. The model was operationalised using a questionnaire and the second, quantitative stage of research entailed the application of the questionnaire instrument using a face-to-face interview method.

The qualitative stage of interviews provided a means of ensuring that the identified beliefs had a strong concordance with the spatial data context that is investigated in this study. Owing to the complexity of the behaviour - spatial data sharing - interviews alone would have been insufficient to provide the necessary information about the full range of beliefs or a basis for translating them into questionnaire items. Therefore, the development of the model of spatial data sharing also employed theoretical insights drawn from several related fields of enquiry.

Structure of the Book

Chapter 2 presents the context for a study of GIS and spatial data sharing beliefs and behaviour. It sets out the technological context by giving an overview of GIS. The chapter also considers the special characteristics of the spatial data that are used in GIS applications and outlines the players involved in spatial data generation and use. The concept of a spatial data infrastructure to co-ordinate spatial data-related collaboration among organisations is presented and an overview of the national, regional and global initiatives world-wide which are grounded in the same basic model (the development of metadata, clearinghouses, core data sets, and standards for data collection) is provided. Analysis of these initiatives demonstrates that they are based on the assumption that key individuals within organisations are highly motivated to engage in spatial data sharing. The only barriers to such sharing are said to be technical ones that can be reduced as a result of agreements. Yet virtually nothing is known about how the motivation to share can be fostered so that inefficiencies in and lost opportunities for the use of spatial data can be reduced. Turning to the country-specific context of the research, South Africa, the institutional set-up for the distribution of spatial data in South Africa is outlined and details are provided about the national spatial data infrastructure in the country.

Chapter 3 introduces the theoretical context for the research. First, the chapter locates this study within innovation studies by relating it to research on the diffusion of innovations. Within the diffusion of innovation studies field, the research considers the conditions under which the different actors involved with spatial data may be willing to share it across organisational boundaries. Second, this chapter reviews the literature that deals specifically with spatial data sharing. Third, it justifies the choice of the TPB (Ajzen, 1991) as a conceptual framework for examining the issue of spatial data sharing by considering alternative decision making models. Finally, the chapter presents the main theoretical constructs of the TPB and outlines the central research questions.

Chapter 4 explains the research design and methods used in this study and how the application of the TPB is applied as an organising framework for the research. The chapter provides a detailed specification of the research design including, first, the choice of the country, South Africa, as a case study within which the empirical research took place; second, a definition of the behavioural elements that are included in the model; and, third, the two stages of the empirical research (qualitative followed by quantitative research) that were employed to analyse the determinants of spatial data sharing behaviour in South Africa. The chapter then presents the methods used to implement these two stages of empirical research. The selection procedure for the interviews, the stratification procedure for the survey sample and the design of the questionnaire instrument are discussed. The chapter also offers some reflections on the strengths and weaknesses of the research methodology.

Chapter 5 presents a model of the factors that were expected to influence the willingness to share spatial data on the basis of the first qualitative phase of the research. Using the TPB as an organising framework, several concepts are developed to provide an integrated model based on the qualitative interview material; insights drawn from the literature on spatial data sharing; and concepts drawn from the research on resource dependence and the knowledge creation process. The resulting model represents the expected structure and content of the determinants of the willingness of individuals within organisations to engage in spatial data sharing. Hypotheses are derived from this model and these are presented together with a discussion about the likely explanatory power of the constructs used in the model. The chapter concludes by considering measures that can be used to assess the actual scope of spatial data sharing on the part of individuals within organisations engaged in GIS activities.

Chapter 6 presents the findings of the survey research and an assessment of the apparent overall willingness of the sample of 112 individuals from 73 organisations in South Africa to engage in spatial data sharing. In this chapter, the results are presented using descriptive statistics. It concludes with an assessment of the actual sharing behaviour by summarising self-reports of actual spatial data sharing by these individuals and their organisations in South Africa.

Chapter 7 presents an analysis of the results of the second phase of the empirical research. This analysis considers which of the elements of the model are most likely to influence the willingness of individuals within organisations to share spatial data and identifies the specific beliefs that appear to influence the willingness of organisations to engage in such behaviour. The practical implications of the analysis are considered by specifying the particular determinants of data sharing that should be taken into account by policy makers. The chapter suggests how efforts to improve the current spatial data sharing initiative in South Africa might encourage greater spatial data sharing.

Finally, Chapter 8 summarises the findings and considers the similarities and differences between the insights generated by the research and elements of the practices and incentives created by current spatial data sharing initiatives by the policy and organisational set-up in South Africa. Conclusions are drawn about the contribution of the application of the TPB to the understanding of the diffusion of the GIS technological innovation and the extent to which it offers a means of examining the social or behavioural aspects of the diffusion process more generally in other contexts of GIS usage and - more broadly still - in studies of the diffusion of innovations in the science and technology policy research field. The chapter reflects on the specific methods used to implement this theoretical approach and concludes by pointing to new areas of research.

Conclusion

The study succeeds in applying the TPB in a way that is revealing in terms of the incentives and disincentives for key individuals within different organisations to engage in spatial data sharing across organisational boundaries. The principal new empirical insight that stands out that could not be gleaned from the existing limited anecdotal studies is as follows. While the technical aspects that are currently the focus of attention in the context of spatial data sharing initiatives - such as the interoperability of different GIS applications and spatial data sets, the establishment of standards, and the implementation of clearinghouses - are necessary elements, it cannot be expected that their resolution will be sufficient to overcome the obstacles to spatial data sharing across organisational boundaries.

The results of this research show that it is important that the current spatial data sharing initiative in South Africa refers to specific incentives that influence sharing rather than only to the general benefits of engaging in spatial data sharing. Moreover, fostering greater spatial data sharing so that GIS can be used more effectively, means the disincentives encountered by individuals (and organisations) also need to be addressed. In particular, it is important to address and reduce the fears that decision makers within organisations associate with losing control over the use of spatial data.

The overall conclusion of the research is that social and behavioural aspects have considerable influence on the specific preconditions for the diffusion of GIS considered in this study, i.e., the availability of spatial data. The prospects for change to address these aspects are promising and lie within the context of emerging spatial data initiatives.

Chapter 2

Spatial Data in Context

Introduction

With the increasing use of GIS, the availability of spatial data is an issue that affects many organisations. They are faced with the high cost and substantial effort involved in the generation of spatial data. The need for greater collaboration and co-operation is being considered increasingly widely in some public and the private sector organisations that play a role in the generation and application of spatial data. Spatial data sharing between organisations is increasingly being discussed as a way of overcoming the obstacles that stand in the way of greater data availability. A wide variety of stakeholders with different interests are involved in national, regional and global spatial data initiatives. These initiatives encompass a variety of technical features and alternatives.

This chapter presents the context for a study of GIS and spatial data sharing beliefs and behaviour. The first section sets out the technological context by giving an overview of GIS. The next considers the special characteristics of the spatial data that are used in GIS applications and the third section outlines the players involved in spatial data generation and use. In the following section, the concept of a spatial data infrastructure to co-ordinate spatial data-related collaboration among organisations is presented. Turning to the country specific context of the study, South Africa, the final section outlines the institutional set-up for the distribution of spatial data in South Africa and provides details about the national spatial data infrastructure in South Africa.

Geographic Information Systems

GIS are information technology applications that depend upon the availability and use of digital spatial data. Originating in the digitisation of maps in the discipline of geography during the 1970s, GIS technology has evolved rapidly over the last two decades and is now being used widely outside the discipline. Nowadays, GIS combine spatial and socio-economic data from numerous different data sets. The applications are intended to support the activities of monitoring, planning, and policy making regarding natural and other types of resources and infrastructures such as transport networks, data networks and waterways. These systems consist of hardware, software and a combination of spatial and socio-economic data from numerous different data sets. The resulting composites of data can add a geographic dimension to development planning because this information can be made visually accessible and located in a context in a way that lends itself to meaningful interpretation from the perspective of the user (Swarts, 1998). The application of GIS is expected to facilitate 'the task of planning and determining priorities in assigning specific resources to specific areas' (Schwabe et al., 1998: 4), for example, by being able to see the exact distribution of facilities and to identify areas with greatest needs. GIS offer the opportunity to render planning activities more effective and efficient by providing structured storage, retrieval and manipulation of information such that an existing task may be performed better, cheaper, or faster, or new problems can be tackled (Longley et al., 1999).

GIS are employed by the public and the private sector in the industrialised countries to achieve a wide range of objectives such as improved town planning, marketing research and environmental monitoring. In these countries, applications are available at comparatively low cost considering the existing levels of skill and equipment for using advanced information and communication technologies together with the trend towards Internet and Intranet-embedded GIS applications and the integration of map information into management information systems. In developing countries, GIS are also increasingly being used by the public and private sectors and by development agencies to achieve improvements in development projects and planning activities. GIS demands for computer processing power have been greatly reduced and personal computer applications are now commonplace. Table 2.1 presents the wide variety of GIS applications utilised for development planning and in development projects.

Table 2.1   Areas of GIS application in development projects

Regional Rural Development / Resource Management
Forestry / Nature Protection
Environment and Resource Protection
Plant Production / Plant Protection / Agrarian Research
Animal Husbandry / Veterinary / Fishery
Water Supply and Waste Management
Education Science
Urban Planning / Urban Development
Food Security Project / Emergency and Refugee Help
Energy and Transport
Agrarian Policy
Health
Irrigation

Source: based on Christiansen *et al.* (1997)

The actual spread of the use of GIS applications is considerable. A recent estimate refers to 461,300 licensed users world-wide with North America accounting for 71%, Europe 14.6%, Asia 6.7%, South America 4.1%, Australia

2.2%, and Africa 1.3% (van Helden, 1999).[3] In terms of economic value, Dataquest (1996) estimated the GIS market world-wide to be worth US$ 2.6 billion in 1995, a 16.5% increase in value over 1994. More recent market research has confirmed a strong growth rate for the GIS market world-wide (Dataquest, 1999).

Although the share of developing country implementations is still far behind that of the industrialised countries, there is evidence that it is growing (Nuttall and Tunstall, 1996; Christiansen *et al.*, 1997). This may be due to the incorporation of these applications in development planning and their use by development agencies for projects with spatial components. The generic nature of GIS with its wide area of applications and the actual spread of GIS in developing countries in recent years strongly suggest that they are applicable in developing country circumstances.

As outlined above, GIS are spreading in the industrialised and the developing countries. Spatial data are a prerequisite for the efficient and effective use of GIS. The growth in the use of GIS is resulting in a surge in demand for digital spatial data. However, a number of characteristics of spatial data can limit their availability, as discussed in the following section.

Spatial Data Characteristics

Spatial data specify the location and characteristics of physical phenomena.[4] Although GIS users generate data particular to their area of application, the operations, nevertheless, depend upon the availability of certain data sets for a wide range of complementary spatial data, as shown in Table 2.2.

Spatial data have a number of characteristics that set them apart from other types of data. For example, the collection process of spatial data sets is complex and expensive, involving a variety of technologies and procedures ranging from ground-based traditional survey methods and photography, to remote sensing, airborne aerial photography and global satellite positioning systems.[5] Furthermore, once collected, spatial data sets need to be maintained and updated or replaced to reflect changes in the physical environment due to, for example, changes in transport infrastructure and urban expansion or the effects of natural disasters.

---

[3] Compared to an estimated 93,000 GIS sites world-wide in 1995 (Burrough and McDonnell, 1998).

[4] Although the terms spatial (or geographic) data and information are frequently used interchangeably both in practice and in the literature, this chapter will refer only to spatial data. Spatial data sets, once used in particular contexts, constitute spatial (or geographic) information.

[5] For example, the US Office of Management and Budget (OMB) estimated that total expenditures on digital geographical information in Federal agencies in 1993 were in excess of US$ 4 billion (Longley *et al.*, 1999).

Table 2.2   Types of spatial data

Geodetic
Land surface elevation / topographic
Bedrock elevation
Digital imagery
Government boundaries /administrative boundaries
Cadastral / land ownership
Transportation / roads
Hydrography / rivers and lakes planimetric
Ocean coastlines
Bathymetry
Physical features / buildings
Place names
Land use / land cover / vegetation
Geology
Real estate price register / land valuation
Land title register
Postal address
Wetlands
Soils
Register of private companies
Gravity network
Zoning and restrictions

Source: based on Onsrud (1999)

Along the spectrum of GIS applications from mapping to development planning, the various applications have fundamentally different requirements in terms of the level of data accuracy and complexity (Abbott 1996; see Figure 2.1).

Land / mapping	Demographic	Socio-economic	Macro-economic
AXIS OF DATA TYPES			

Increasing detail

ACCURACY OF SPATIAL REFERENCING REQUIRED

Increasing complexity

COMPLEXITY OF DATA

Figure 2.1   Accuracy and complexity of data
Source: Abbott (1996: 9)

This results in accuracy and incompatibility problems such as locational uncertainty owing to differing geographical referencing systems; scale differences and map-scale-dependent accuracy and resolution; different degrees of generalisation; boundary and location data errors; differing timing of enquiries; varying reliability of data (Coppock and Rhind; 1991; Openshaw *et al.*, 1991; Flowerdew and Green, 1991). The scale and error characteristics of a spatial data set, therefore, need to be well documented to ensure that the data are relied upon in a way that is consistent with the reliability established by the originator. Such data about spatial data, or 'metadata', are particularly important since GIS software can hide the level of uncertainty about the data from the unacquainted user.

Similar to other types of data, but more crucially so, spatial data need to be up-to-date, reliable, and affordable, and specific skills and financial resources are required to collect, acquire, integrate and update the data sets.

Spatial Data Players

In parallel with the increase in public and private sector users of GIS, most national mapping agencies that traditionally were involved in surveying and mapping national territory for military and government planning purposes have now digitised their operations. They are major suppliers of national spatial data sets on differing scales. In some developing countries such as Mexico and South Africa, for example, the national mapping agencies, the National Statistics, Geography and Informatics Institute (INEGI) and the Chief Directorate for Surveys and Mapping, respectively, have made considerable investments in the digitisation of their map inventories and are using advanced geographic information technologies (Jarque, 1997; Clarke, 1997).[6] Due to the wide area of application, however, the demand for different types of data sets cannot be met by these organisations alone.

Aside from the national mapping agencies, many users such as oil companies or the utilities with sufficient resources and spatial data capturing skills engage in the collection as well as the use of spatial data. They may even make some or all of their spatial data sets available outside their organisations to recover some portion of their investment. Thus, they act both as users and suppliers of geographic information. Finally, commercial organisations add value to data sets by integrating and then selling existing data sets (Burrough and McDonnell, 1998).

---

[6] For example, the Mexican National Geographic Information Systems (MNGIS) comprises capture, production, organisation, integration, analysis and presentation processes. The modernisation programme has been carried out in each of the 10 regional offices of INEGI in Mexico. The system includes advanced software and equipment, and 73 specialists were trained over a period of 9 months (Jarque, 1997).

Spatial Data Infrastructure Initiatives

The broader context for collaboration and the distribution of spatial data within which individual GIS implementations function is illustrated by a variety of initiatives that aim to facilitate spatial data sharing. While the Open GIS Consortium (OGC) of public and private sector actors is primarily concerned with the technical standards for spatial data transfer between different GIS software vendors (OGC, 1999), the spatial data infrastructure initiatives that are emerging at the national level in industrialised as well as developing countries (and at the regional and global levels) are more comprehensive.[7] As summarised in the Executive Order of the President of the United States, a national spatial data initiative involves: '... the technology, policies, standards, and human resources necessary to acquire, process, store, distribute and improve the utilization of geospatial data' (Federal Register, 1994: Section 1(a)).

Based on a world-wide survey (Onsrud, 1999), these initiatives typically encompass the following components;

* *Core data sets*
* *Metadata*
* *Data standards*
* *Clearinghouse*

These initiatives generally are striving to arrive, for example, at agreed pricing structures and policies, core data sets that may or may not be free of charge, copyright agreements and statutory requirements for metadata acceptable to the many different actors. The core data are a reflection of the realisation that GIS users from all areas of application require access to certain common spatial data sets and, most fundamentally, a common co-ordinate system to which other data can be referenced. Considering these core spatial data sets as a national resource in support of GIS implementations, the model for their provision is being debated and varies considerably between nations ranging from merely charging for the cost of

---

[7] The survey by Onsrud (1999) received responses about their national spatial data infrastructures from Antarctica, Australia, Canada, Columbia, Finland, France, Germany, Greece, Hungary, India, Indonesia, Japan, Kiribati, Macau, Malaysia, The Netherlands, New Zealand, Northern Ireland, Pakistan, Russian Federation, South Africa, Sweden, United Kingdom, United States. Regional initiatives include those in Australia and New Zealand, Asia and the Pacific, and Europe. Recently, INSPIRE (The INfrastructure for SPatial InfoRmation in Europe initiative) has been launched as a regional initiative at the European level for the EU Member States and Accession Countries.

In addition, initiatives to develop spatial data infrastructures are underway at the national level in Zambia, Ghana, and Botswana, and at the regional level: the EIS (Environmental Information Systems) Program; activities by the SADC Regional Remote Sensing Unity; CODI-GEO (Sub-committee on Geo-information of the Committee on Development Information by UNECA); AFRICAGIS conferences and exhibitions; and an Africa-wide SDI (Kangethe, 1999).

At the global level, regular conferences concerned with a global spatial data infrastructure have developed into an umbrella organisation for national and regional initiatives called Global Spatial Data Infrastructure (GSDI).

data distribution, to full recovery of all costs of generation and distribution, to profit making.[8]

At the technical level, the initiatives aim to arrive at compatible standards such as those for spatial data collection and exchange, and metadata. Standardised documentation about a spatial data set is collected as metadata, describing the characteristics and detailing the ownership of the spatial data set. Another element is the necessary standards for the collection and exchange of spatial data sets that need to be agreed upon by, and applied to, the users, producers and value-adding organisations. Finally, clearinghouses are a central part of the initiatives. They are envisaged, and in many cases are already implemented, to act as focal points for identifying and accessing spatial data sets that are of interest to users other than those who have generated the spatial data. The clearinghouse provides a database of metadata that typically can be accessed via the Internet and, in some cases, on CD-ROMs. The components are pictured in Figure 2.2.

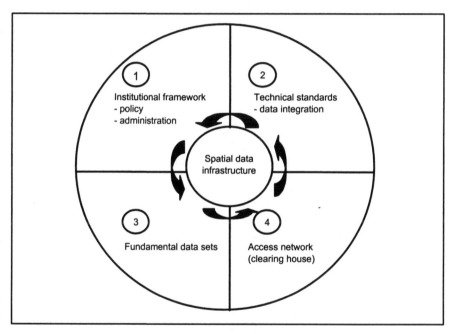

Figure 2.2   A model of a spatial data infrastructure
Source: CIE (2000: 11)

The impetus for these initiatives comes predominantly from governments since public sector agencies are major collectors, providers and users of information that

---

[8] Full cost recovery refers to making spatial data available at a price that allows the originator to recover all the costs incurred in producing the data set. Marginal cost of reproduction entails charging merely for the time and medium (e.g. CD-Rom) required for providing a copy of the spatial data (sub)set.

has spatial components (Burrough and McDonnell, 1998). Some examples of the use of this information in developing countries are to produce social service indicators such as crime (density of crime events), health (urban and rural fertility rates), and school needs (sanitation, learner/teacher ratio, electricity supply). The role of a national spatial data infrastructure in a developing country is illustrated in the next section focusing on the evolution of the National Spatial Information Framework (NSIF) in South Africa. This discussion is used to show that, even when the common components of a spatial data initiative are explicitly addressed, this is not sufficient to produce a process that leads to the desired levels of collaboration and exchange aspired to by the proponents of these initiatives.

Spatial Data Distribution in South Africa

In South Africa, considerable GIS investments have been made by the public sector. GIS are used widely in the government sector at the national, provincial and municipal levels (van Helden, 2000), by para-statal organisations such as ESKOM (electricity), TELKOM (telecommunications), CSIR (scientific and industrial research), and by academics, and the private sector is quickly catching up.[9] A particularly prominent example of a GIS application is the use by the Independent Electoral Commission in collaboration with the Chief Directorate for Surveys and Mapping, the Surveyor General and Statistics South Africa. Combining demographic information (census data), with topographical (geographic position of rivers, roads and contours) and cadastral data (land parcel delimitations and other man-made boundaries) in this GIS, electoral wards and voting districts were demarcated for the 1999 elections (Lester, 1999; Martin, 1999). Howell (1999) suggests that considerable financial savings were attained by knowing where people would vote in the 1999 General Election and that this information was helpful in reducing the time voters had to queue at the voting stations.[10] Yet it was not clear when, to whom and at what cost the Independent Electoral Commission's value-added spatial data sets would be available after the elections.

Spatial information management initially received attention in 1988 with the establishment of the National Land Information System (NLIS) facilitated by the Chief Directorate for Surveys and Land Information, the national mapping agency in South Africa. The use of GIS in the public sector has become increasingly pervasive. Contributions of GIS to the reconstruction and development process in South Africa include support of the process of land reform, for example, by facilitating participatory and community driven planning for the allocation of low cost housing,[11] and resource allocation, for example, by targeting funding for

---

[9] Personal conversation with the two market-leading GIS software suppliers in South Africa (Computer Foundation, GIMS Ltd.), March 1999.

[10] For example, ZAR 34 million in staff costs, ZAR 9 million for ballot papers (Howell, 1999).

[11] An participatory GIS is envisaged as '... a GIS application where local knowledge, community needs, and specific social histories are appreciated and incorporated into the

nutrition programmes away from urban centres to needy rural communities based on a number of selection criteria[12] (MacDevette *et al.*, 1999). Owing to the diffusion of GIS in the public sector, the Reconstruction and Development Programme (RDP) Ministry in the Office of the President commissioned a report in 1996 to consider the co-ordination of the use of GIS for development planning in the public sector at the national level based on interviews with government and provincial departments and para-statal organisations (Abbott, 1996). The report reviewed a number of national and provincial GIS initiatives linked to development planning that were underway at the time and concluded that the distribution of spatial data was a crucial issue for the co-ordination of GIS.

Subsequently, in 1997, the Department of Land Affairs created a directorate for the NSIF which was endorsed, along with a Committee for Spatial Information, by the Cabinet in 1999.

The National Spatial Information Framework in South Africa

The NSIF in South Africa is regarded as an infrastructure that needs to be provided and maintained in order to pursue social and economic goals (Gavin, 1998). The specific aims of the NSIF are summarised in the following quote:[13]

---

development process ...' (Harris et al., 1995: 197). However, the debate within geography regarding GIS, power and knowledge has argued that participatory policy formulation using GIS is still facing a number of problems (Harris et al., 1995; Shiffer, 1999). Hence, '... there is still the need for much participatory research and development around GIS before communities are likely to be able to use the full power of the technology.' (MacDevette et al., 1999: 922)

[12] Geographical locations were identified on the basis of the following criteria: the highest percentage of poor people, a high number of dependants per person, and high levels of female illiteracy (MacDevette *et al.*, 1999).

[13] A number of efforts at establishing, co-ordinating and aligning spatial data exchange and distribution were already underway in the public and the private sector in the region before the establishment of the NSIF. For example, prior to its inception, the joint public/private sector initiative, Southern African Metadata consortium (SAM), by CSIR, Geographic Information Systems (GIMS) and Integrated Spatial Solutions (ISS) was already implementing a clearinghouse for the metadata of participating organisations (Smith, 1998). A second initiative was run by the South African National Defence Force (SANDF). The spatial data used by the military originates from the South African Engineer Formation (landward mapping and aeronautical charts), the Navy (nautical charts) and the Air Force (photography and imagery). The SANDF maintains a directorate for the co-ordination of spatial data exchanges with the support of a database of metadata. It uses the Digital Geographic Information Exchange Standard (DIGEST) set of standards, originating from NATO countries for military applications, for the exchange of spatial information among users and producers (Fourie, 1998). However, neither of these two initiatives parallels the scope of the NSIF. The NSIF has built on the technical achievements of the other initiatives by purchasing their products (e.g. metadata database software).

The aim of the NSIF is to expedite the distribution and facilitate the integration of disparate geographic data sets in order to enable planners to make sound development decisions based on accurate, complete and current geographic information. Further, the NSIF will also increase efficiency and savings in government by eliminating duplication in the capturing, storing and maintenance of geographic data sets. (Gavin, 1998: 18)

The NSIF endorses the typical elements of the national spatial data infrastructures identified above, that is, a catalogue of metadata, an Internet-enabled clearinghouse facility, standards for metadata and spatial data collection, and the identification of a number of core data sets. Explicit attention was to be given to GIS capacity building through the NSIF that would allow the identification of people with the necessary expertise and skills that could of assistance and share their experience. Efforts were underway to develop policies (on copyright, pricing of data sets, and liability and responsibility of data providers) and standards (metadata and data exchange standards, and classification and co-ordinate system standards) that might serve to promote data sharing (NSIF, 1998a). A comprehensive list of interim policy guidelines on access principles, custodianship, privacy principles and pricing was made available (NSIF, 1998b).

The NSIF directorate recognised that in order to meet its objectives, the involvement of non-governmental actors, such as private sector, academic and para-statal stakeholders was important. A number of workshops with public and private sector, para-statal and academic institutions was held throughout 1998 to address specific components of the NSIF and representatives from all stakeholder spheres were encouraged to participate in the policy and standards task teams.

Yet the initiatives of the NSIF have focused primarily on the potential public sector benefits:

... the primary goal of achieving a co-ordinating system and procedure in the capture and management of spatial information through the NSIF is ensuring that information is available and can be utilized for development planning by government. (NSIF, 1998a: 6)

In line with this reasoning, only public sector representatives were included in the National Committee which was in charge of the NSIF. This restriction was put in place because the NSIF's focus on spatial information forms part of a larger initiative to facilitate effective information management in government, called the Government Information Project (GIP). As a result, the development of the NSIF has been dominated by the public sector; yet the NSIF is expected to be endorsed by the private sector as well. Thus it appears that the scope of the NSIF was not clearly defined.

At the policy making level, the NSIF may be facing difficulties in implementing the national spatial data infrastructure due to its position in the Department of Land Affairs. In the other two directorates of this national government department there are two of the most important spatial data suppliers - the Chief Directorate of Surveys and Mapping (which supplies the topographical data sets) and the Chief Surveyor General (which maintains the cadastral data base). The policy making capacity of the NSIF with regard to such crucial issues as the provision of core data sets and their pricing policies is therefore limited.

This illustration of the evolution of the NSIF in South Africa demonstrates that there is a wide variety of technical, economic, political and social issues at stake concerning the distribution of spatial data. The actual emphasis in the NSIF is on the implementation of clearinghouses, standards, core data sets, and pricing and copyright policies. However, underlying the NSIF, there is a need to encourage a shared vision and behaviour among a group of actors, i.e. the GIS community. This would entail creating an environment in which individuals in public and private sector organisations dealing with geographic information will actually wish to participate in data sharing across organisational boundaries. For example, Pienaar and van Brakel (1999) have suggested that co-operation among different organisations is a crucial aspect in order to benefit from the establishment of clearinghouses on the Internet:

> For the GIS community in Africa and Southern Africa to benefit from sharing GIS datasets via the web, improved cooperation is needed. The web as a medium can undoubtedly be seen as a breakthrough in spatial data sharing, but is dependent on organisations to put some effort into making data available for sharing. (Pienaar and van Brakel, 1999: 370-371)

However, organisations that are not dependent on others for spatial data may perceive that they have little to gain by engaging in spatial data sharing activities with other organisations.

The South African NSIF initiative recognises the need for a spatial data sharing 'culture' (Clarke *et al.*, 1998). Although an important step, those responsible for the NSIF have not elucidated or addressed whether the actual motivation of individual actors within organisations favours participating in, or abstaining from, spatial data sharing. For example, the interim policy guidelines merely state that:

> Organizations should work together to facilitate data and information sharing and avoid duplication of data capture and cost to Government. (NSIF, 1998b :1)

The social network of users, producers and value-added organisations together with the workshops that were being held provided opportunities for participants to voice their concerns about spatial data availability and sharing. At these workshops, the organisers placed emphasis upon identifying the spatial data 'needs' of the users, the 'supply' of spatial data sets of the producers and matching them, effectively reducing their efforts to facilitating a 'market place' between these actors. This approach leaves the underlying willingness and capability to participate in spatial data sharing outside the forum for debate. Yet the successful use of GIS is reliant upon the willingness of organisations to make their data accessible across organisational boundaries, rather than confining data access to organisational boundaries, so as to avoid duplication, inefficiency and lost opportunities for the use of spatial data to assist in solving problems.

Similarly, initiatives at the national, regional, and global levels also have failed to address the willingness of individuals within organisations to engage in data sharing, assuming, instead, that the main considerations of the actors involved are the reduction of costs and duplication of effort.

The role of spatial data initiatives entails not just the co-ordination and development of metadata, clearinghouses, and standards components. The important aspect that may be expected to vary considerably for each national initiative is the existence, or fostering, of what has been referred to as a data sharing 'culture' (Clarke *et al.*, 1998) that encourages actors to participate in spatial data sharing.

Conclusion

This chapter has introduced the characteristics of spatial data and the range of actors involved in using, producing and adding value to spatial data. As the number of GIS users is increasing in developing and industrialised countries and they are facing the high cost and considerable effort involved in the generation of spatial data, the need for collaboration and co-operation is claiming the attention of increasing numbers of individuals within some public and the private sector organisations. Numerous similar initiatives are underway at the national, regional and global levels. Although the necessary standards, procedures and policies dealing with the technical, economic and political aspects of spatial data interchanges are being addressed, the promoters of the spatial data infrastructure initiatives do not appear to acknowledge that, in order for their initiatives to be effective, the determinants of spatial data sharing behaviour need to be explicitly considered, so that important non-technical barriers that are likely to discourage data sharing can be overcome.

In order to investigate whether spatial data sharing behaviour is likely to follow from initiatives that are underway to promote standard setting, the establishment of clearinghouses and consistent pricing policies for the national spatial data infrastructures, it is necessary to identify the existing incentives and disincentives to participate in data sharing that are present within the communities of actors and their interactions with respect to the generation and use of spatial data.

Chapter 3

A Behavioural Approach to Spatial Data Sharing

Introduction

This chapter considers the issue of spatial data sharing across organisational boundaries in more detail and provides the theoretical context for this research. The overall argument is that the issue of spatial data sharing needs to be addressed by explicitly taking into account the social factors that may influence data sharing. The resultant insights are expected to aid the promotion of spatial data sharing across organisational boundaries to help GIS to be used more effectively. This study examines the nature of the conditions under which the different actors involved with spatial data are willing to engage in sharing it.

The first section locates the research within the field of innovation studies by relating it to research on the diffusion of innovations. Next, the literature that deals specifically with spatial data sharing is reviewed. The third section justifies the selection of the theoretical approach of this study, the Theory of Planned Behaviour, by considering alternative decision making models. The final section introduces the principles of the TPB and outlines the central research questions.

Diffusion of Innovation Studies

A significant amount of research within the field of innovation studies is concerned with measuring and explaining the sources, determinants, and socio-economic implications of technical change. Much of this work concentrates on the effects of technical change as reflected in increased productivity and in the rate of economic growth to examine the connection between technology and economic progress. The diffusion of innovations constitutes an important aspect of technical change because it is only when innovations are spreading that they can 'give rise to significant economy-wide effects on the pattern of investment and employment' (Freeman, 1982: 214).

An important framework for diffusion research was developed by Rogers in 1968 and updated recently (Rogers 1995). Diffusion research seeks to explain how and why an innovation spreads within a given group of adopters and different models have been proposed that are concerned with how the mechanics of the diffusion process can be best described (for a detailed review, see, for example, Sarkar, 1998). However, the reasons why an innovation is not diffused or may even be rejected are less often addressed. This study considers the preconditions for the diffusion of a technology such that its users will have a greater likelihood to behave in ways that may increase the potential benefits of the technological application.

The specific innovation considered in this research is GIS. Spatial data are indispensable for the use of GIS and national, regional, and global initiatives have been developed to promote spatial data sharing as a way of extending the availability of spatial data. The aim of this research is to examine the conditions under which the different actors involved with spatial data may be willing (or unwilling) to share these data across organisational boundaries.

Spatial Data Sharing: The Current State of Understanding

As Chapter 2 has highlighted, an important role of spatial data initiatives involves not just the co-ordination and development of metadata, clearinghouses, and standards components. The important aspect that can be expected to vary considerably for each national initiative is the existence or the fostering of a data 'sharing culture' (Clarke *et al.*, 1998) that encourages actors to participate in spatial data sharing. A growing body of literature on GIS has explored the diffusion of GIS (e.g., Campbell, 1991; Onsrud and Pinto, 1991; Masser and Onsrud, 1993; Rogers, 1993;), the evaluation of GIS implementations (e.g., Aangenbrug, 1991; Taylor and Johnston, 1995; Nedovic-Budic 1998), the social and ethical implications of GIS (e.g., Pickles, 1995; Curry, 1995; Martin, 1996), the organisational context of GIS (e.g., Campbell, 1996; Harvey and Chrisman, 1998), and the usefulness of GIS for planning (e.g., Webster, 1994; Heikkila, 1998) and for development (e.g., Dale and McLaughlin, 1988; Taylor, 1991; Harris *et al.*, 1995; Klostermann, 1995; Clarke, 1997; Jarque, 1997).

The growing importance of spatial data availability and data access for the effective use of GIS has begun to generate a branch of research that focuses directly on the issue of spatial data sharing. This body of work was consulted to develop a definition of spatial data sharing for the purpose of the present study. This definition is used as a basis for examining whether and why individuals within organisations may or may not wish to engage in spatial data sharing with other organisations.

Conceptualisations of Spatial Data Sharing

Spatial data sharing has been defined in a number of ways. These definitions are summarised in Table 3.1 in order to clarify the commonalties.

Table 3.1   Summary of spatial data sharing definitions

Spatial data sharing is defined as the (normally) electronic transfer of spatial data/information between two or more organizational units where there is independence between the holder of the data and the prospective user. The transfer may be in periodic bulk-transfer, routine daily transfers, or on-line access driven by individual transactions. The participants must be separate organisations or may be departments within the same organisation. For our purpose, the distinguishing characteristic of spatial data sharing is that there be an arm's length exchange or transfer. (Calkins and Weatherbe, 1995: 66)

Sharing of spatial data implies an ongoing relationship between the two participants. (Calkins and Weatherbe, 1995: 70)

The sharing of geographic information may take many forms, ranging from the sale of data by one organization to another to simultaneous access of a single data set by many persons or organisations. (Kevany, 1995: 77)

Data sharing implies an interaction between individuals or among organisations comprised of individuals. (Tosta, 1995: 201-2)

Sharing of geographic information involves more than simple data exchange. To facilitate sharing, the GIS research and user communities must deal with both the technical and institutional aspects of collecting, structuring, analyzing, presenting, disseminating, integrating, and maintaining spatial data. (Onsrud and Rushton, 1992: 1)

Data sharing presupposes a strong, long-term, funded commitment to reconcile what really constitutes different versions of the same reality. (Sperling, 1995: 391)

'... process of spatial data interchange' (Sperling, 1995: 394)

Sharing of geographic information necessarily presupposes the existence of relationships among individuals, organizations and / or governmental unit. As spatial databases are developed that are accessible to others than those who created them, relationships among the involved parties and societal control over the databases will be adjusted through legal mechanisms. (Onsrud, 1995: 293)

Data integration [is] the ability to share access to data sources or access common databases. (Dueker and Vanra 1995: 153)

The appropriate focus for sharing data (and systems) resources is data and systems integration. Systems integration is a means by which data sharing can be achieved, and data integration is a compelling reason for sharing data. Integrating or combining data in a GIS increases its effectiveness and creates opportunities for wider enterprise benefits that accrue to entire organisations and constituencies. (Dueker and Vanra, 1995: 169)

The 'data sharing' issue is much more complicated than simply determining how data created by one organization can be used by other organizations. The ability of different organizations to cooperate will determine what data is available and what technology is used. (Bamberger, 1995: 136)

Table 3.1   continued

'... inter-organizational information sharing is an activity that occurs within the framework of an inter-organizational alliance. Such alliances can result from a number of pre-conditions ... one organisation may hold power over another and can therefore force the subordinate organization to cooperate ... appeals to professionalism and common goals motivate organizations to ally voluntarily ... in many instances, organizations of more or less equal power status negotiate with one another to achieve an acceptable arrangement for inter-organizational cooperation.' (Obermeyer, 1995: 139)

'... spatial data sharing is in essence a problem of inventing and building a "spatial information infrastructure" of people, organizations, and technology on top of today's GIS tools, data, and operations.' (Evans and Ferreira, 1995: 454)

'... the reliability, accessibility, and feasibility of data sharing can depend on whether data are shared among multiple departments that depend upon the data to feed a divergent set of multi-user applications, or whether data are only informally shared among one department's technically able staff. Interchange can be highly structured, via formal requests, or ad hoc, on a "help yourself" basis ...' (Evans and Ferreira, 1995: 449)

Effective sharing of geographic data on a widespread basis requires that individuals and organizations must be able to determine whether data has already been collected by others and whether that data is fit for their proposed use. The ability to access and acquire the data efficiently must also be provided. A spatial data infrastructure may be defined as the distribution system that enables effective access to digital geographic data, the coordinating and control structures that develop and maintain data sets, and the geographic data sets accessible through the system. (Onsrud and Rushton 1995: 500-501)

'... organizational information sharing and cooperation are conceptualized as the degree, extent, and nature of interpersonal relationships among members from multiple organizations.' (Pinto and Onsrud, 1995: 58)

'...coordinated development and use of local geographic information systems and databases.' (Nedovic-Budic and Pinto 1999a:183)

'...interorganizational development and sharing of geographic information systems (GIS and databases ...multi-participant GIS activities...' (Nedovic-Budic and Pinto 1999b:53)

Information sharing also happens among organizations that are not necessarily independent. For example, organizations get involved in joint data acquisition and database development for a common function or a project, or are engaged in building, maintaining, and using interorganizational system. (Nedovic-Budic and Pinto 2000: 456)

'... organizations may create a spatial database useful to several other agencies ... This is the typical case where "data sharing" is appropriate since such efforts are costly but, if done with consideration of needs and standards of other agencies, present an opportunity to save money and time for all participants through cooperation ... Some spatial data sets, however, are so broadly useful that they deserve not just to be "shared" but to be recognised at all levels of government as the "foundation of the economy in the 1990s" ... Such recognition should extend beyond "sharing" considerations to making a set of critical spatial sets as accessible and easy to use by analysts as our present highway system is for motorist.' (Cooke, 1995: 366)

The commonalities of these definitions are, firstly, that they highlight the importance of the interactions, interchanges, and relationships between the actors engaged in spatial data sharing. Secondly, spatial data sharing is conceived to take place in a variety of forms and under a range of different terms and conditions. Thirdly, the principal aspect of spatial data sharing is *access* to spatial data.

Yet the use of the term 'sharing' is limiting since it does not intrinsically take into account the two-way dimension of the interaction between the parties involved, be it within partnerships or as part of national spatial data initiatives. Onsrud (1995) has advocated the use of the term spatial data 'sharing' in favour of 'exchange' because the information is still available even after it has been made available to others. However, to 'share' indicates only a one-way relationship, in this case the data flow, but this conceptualisation poses problems when the aim is to try to capture other aspects that are involved such as financial payment or payment in kind for the data.[14] Conceptualising this phenomenon, instead, as an exchange implies a two-way transfer between the participating parties and permits consideration of the characteristics of the two-way flows and the types of relationship that may exist between the parties involved.[15] There are many activities which are related to, and act as facilitators for, the exchange of spatial data. The principal underlying activity, however, is that of making spatial data accessible to, or from, other parties, and this may be expected to take place under certain conditions and not under others. In the light of these clarifications, the following definition of spatial data sharing is used for this research: *Spatial data sharing entails making the digital spatial data used in Geographic Information Systems (GIS) accessible to, or from, other parties. These exchanges may or may not include barter, financial payment or payment in kind.*

The term 'spatial data sharing', commonly used in the GIS community, is clarified to refer more precisely to an exchange relationship in which access to the data is enabled under certain terms and conditions.

In attempts to capture the various conditions that may facilitate spatial data sharing, different classifications have been proposed focusing on access, exchange dynamics and conditions, and on the frequency and scope of sharing (summarised in Table 3.2).

A comprehensive taxonomy of the characteristics of spatial data exchanges has also been developed by Calkins and Weatherbe (1995) which enables recognition of particular types of occurrences of spatial data sharing. This classification of the characteristics of spatial data sharing provided important elements for the empirical part of this study (see Chapter 5) such as the magnitude of spatial data exchanges, who is sharing with whom and the terms and conditions under which spatial data are shared.

---

[14] Dictionary definition (Webster 1960) 'To share: 1. To divide and distribute in portions; to apportion; divide. 2. To partake of, use, experience, or enjoy with others.'

[15] Dictionary definition (Webster 1960) 'To exchange:1. To part with, give, or transfer to another for an equivalent; specifically, to obtain or to supply something else in place of (goods returned). 2. To part with for a substitute, as a palace for a cell. 3. To give and receive or lose and take reciprocally, as things of the same kind; to barter; swap.'

Table 3.2   Classifications of spatial data sharing

Focus	Classification	Author(s)
Access	*Different levels of access to spatial data* King (1995: 264-5) - Access to information from the database (access to only part of the information stored in the database, request for a specific maps (layers of information) for specific locations (cities, areas) - Access to all of certain kinds of information in the database (i.e. data sets) - Access to the database itself (a copy of the actual database)	
Access and distribution	*Classes of sharing* Kevany (1995: 81-2) - Direct access to a single database that is shared among multiple organisations. - Joint compilation or acquisition of data by multiple organizations, each of which obtains a separate copy of the data. - Acquisition of a copy of one organization's data by another organization on a one-time or periodic basis. - Sale of copies of data by an organization to other organizations. - Distribution of copies of maps or geographic data reports by one organisation to other organisations.	
Exchange dynamic and conditions	*Typology of multi-participant GIS setting sharing a geographic database:* Azad and Wiggins (1995: 25-26) - An organization undertakes to provide geographic databases to other organisations at a nominal charge that does not reflect the production costs, or in a 'one-shot' project where maintenance is not a concern. The dominant inter-organizational dynamic here is 'one-way' (i.e., the organization providing the database is not affected by the user organizations wanting or not wanting to take part in the production of the database). - An organization undertakes to provide geographic databases of universal value to a variety of other organizations. Continued maintenance and expansion of the database depends to a large degree on the using organizations picking up a portion of the tab for this effort as well as the demonstration of the usefulness of this data to resource-providing bodies. Moderated 'one-way' dynamic between organizations with demonstrated demand for the product by user organizations. - Several organizations undertake building and maintaining geographic databases by sharing the costs and products as a response to resource scarcity and to minimise duplication/ redundancy. 'Two-way' dynamic reliant on mutual consent and participation of each organisation.	
Frequency and scope	*Inter-organizational information sharing context:* Pinto and Onsrud (1995: 46) - Non-routine, non-recurring: Situation-specific, project driven, work on a common problem, data and expertise shared to solve the problem. - Case-by-case or long-term data sharing: Need for similar information by different agencies addressing different problems, develop procedures for regular sharing and exchange of information but protocol inappropriate for sharing with broader community or additional parties. - Routine sharing process: Standardised, generalisable pattern of exchange: info readily accessible to all parties in terms of its location and format.	

Research into Spatial Data Sharing

Several efforts have been made to gain an understanding of why organisations may or may not wish to engage in spatial data sharing. The factors that have been considered are mostly conceptualised in terms of facilitators and constraints (Tosta, 1995; Stage, 1995), costs and benefits (Dueker and Vrana, 1995; Dangermond, 1995; Alfelor, 1995) or antecedents and consequences (Obermeyer and Pinto, 1994; Pinto and Onsrud, 1995) of spatial data sharing.

Obermeyer and Pinto (1994) have proposed a conceptual framework for spatial data sharing across organisational boundaries. This model of inter-organisational sharing combines a number of variables as facilitators and inhibitors, and as outcomes. These variables consist of the antecedents that may improve the likelihood of creating positive, collaborative relationships between organisations and of the consequences of spatial data sharing. Although Obermeyer and Pinto argue that the variables in the model have been drawn from organisation theory, intergroup dynamics, exchange theory, and political-economy, a major limitation of this framework is the lack of any justification for the inclusion of these factors in the model and for the exclusion of other factors.

Kevany (1995) has proposed a structure for observing spatial data sharing by exploring the factors and conditions that may create a environment conducive to sharing. These factors consist of: a) sharing classes, b) project environment, c) need for shared data, d) opportunity to share data, e) willingness to share data, f) incentive to share data, g) impediments to sharing, h) technical capability to sharing, i) resources for sharing. A number of measures for each of the factors is proposed in order to determine the probability of successful sharing. Although very comprehensive, Kevany's method of analysing the environments for spatial data sharing has several limitations. There is no explicit basis for the inclusion of the factors, nor is there a basis for conceptualising the likely relationships between these factors. The factors are derived on the basis of the author's personal insights and experiences with GIS design, implementation and operation projects. Hence, there is no organised framework in which to analyse the relevance of these factors. Moreover, the framework proposed by Kevany has not been verified empirically.

Azad and Wiggins (1995) have proposed a research framework that focuses on inter-organisational dynamics. The central tenet of their framework is autonomy. Azad and Wiggins assume that sharing results in a loss of autonomy and increased interdependence of organisations and they argue that these consequences are at odds with the organisational goals. These are the principal obstacles to spatial data sharing that have to be addressed. Their framework consists of a typology of inter-organisational spatial data sharing and the reasons for the development of inter-organisational relationships (see Table 3.3).

Table 3.3 Fundamental reasons for geographic data sharing

Type	Necessity	Asymmetry	Reciprocity	Efficiency	Stability	Legitimacy
				Reasons		
I	Legal decision requires one-shot analysis	Organization wants to control information as sole source	Source organization has resources available; other parties' resources not relevant	Source organization sees GIS as providing efficiency gains in its own mission	Source organization has stable mandate, little environmental uncertainty	Source organization wants to improve quality of its own data collection and analysis
II	Legislative mandate encourages initiation of GIS	Lead organization wants to retain control of data quality and standards	Continuing maintenance and improvement of data over time requires more resources than lead agency has	There is a desire to limit data collection redundancy	Lead organization experiences some funding uncertainties in the future	Lead organization wishes to gain more visibility and credibility
III	State mandate for GIS effort	No single organization is able to act or fund a GIS unilaterally	Joining together is the only way a GIS can be implemented	Each participant sees individual efficiency gain for its own organizational goals	Most of the participants are faced with funding uncertainties	Participants wish to gain higher visibility through the GIS

Note:
Type I: one-off provision of spatial data at nominal charge, maintenance is not a concern, organisation providing spatial data is not affected by receiver organisation wanting or not wanting to take part in the production of the database, 'one-way' dynamic.
Type II: organisation undertakes to provide spatial data of universal value to a variety of other organisations, continued maintenance and expansion of the database depends on using organisations' funding, 'one-way' dynamic moderated by demand for the product by user organisations.
Type III: several organisations undertake building and maintaining spatial databases by sharing costs and products, 'two-way' dynamic, depending on mutual consent and participation.
Source: Azad and Wiggins (1995: 30) © Center for Urban Policy Research, Rutgers University

In addition, Azad and Wiggins (1995) argue that the likelihood of establishing inter-organisational relations is related to the intensity of the relationship and the loss of organisational autonomy required by the relationship. Finally, they propose a process model to manage the development of inter-organisational relations for spatial data sharing.

The starting point for the Azad and Wiggins framework is open to question. The assumptions that, firstly, spatial data sharing necessarily leads to a loss of autonomy and increasing interdependence among organisations and that, secondly, these consequences are inherently negative from an organisation's point of view have not been subject to empirical confirmation. Similar assumptions about the nature of the power of, and control over, spatial data are embedded in much of the literature as indicated by the following:

> Because geographic information has potential value to those with effective access to it, this realization gives rise to the desire to exercise ownership rights over the information. Thus, the power that information provides is antipathetic to sharing. (Onsrud 1995: 293)

> If we agree that the possession of [spatial] information serves as a source of control for individuals and organizations, then we are faced with questions about the ways in which organizations can be induced to relinquish this control. (Obermeyer and Pinto, 1994: 107)

In contrast, in this study it is argued that it is not possible to presume the obstacles to sharing *a priori*. Instead, they should be the subject of empirical verification. As Pinto and Onsrud (1995) have suggested that:

> ... little is known, for instance, about the reasons why governmental agencies and other GIS-using organizations will or will not share GIS-related information. (Pinto and Onsrud, 1995: 48)

The point of departure for research into whether and why individuals within organisations may be willing to engage in spatial data sharing should be located one step back owing to '... the potentially differing perceptions of the benefits from [spatial] data sharing' (Sperling 1995: 391).

Perceptions may also vary with respect to the costs of, or obstacles to, sharing spatial data. The nature of these perceptions should not be generalised across organisations without recourse to an empirical investigation that permits validation of some of these assumptions and that provides a basis for generalisation. Rhind (1998) has provided a summary of the views of different players and stakeholders in spatial data and their agendas (see Table 3.4).

Table 3.4 **Possible agendas of various players and stakeholders in spatial data**

'Player'	Alternative short term agendas
Central Government organisations	■ avoid expense – 'hide data'? or ■ maximise use inside and outside department or ■ maximise revenues and minimise costs subject to equality of treatment and fairness
Local Government	■ avoid expenses – 'hide data'? or ■ maximise use inside and outside department or ■ maximise revenues and minimise costs plus conform with national requirements for administrative/statistical data
Commercial sector - information trader / publisher	■ trade profitability and have positive cash flow ■ minimise costs of getting data from elsewhere ■ minimise risks by pre-publication agreements to purchase ■ disaggregate markets and appropriate as much as possible for customer value, prevent arbitrage ■ projectise all activities to measure costs/ benefits/ write-offs
Commercial sector conglomerate	■ if focused as one business, 'gestalt' aspects of business (e.g. data helps sell equipment) dictate decision-making and agenda. ■ more normally, individual enterprises within the conglomerate are judged first on their own 'bottom' lines
Non-Governmental / Not for Profit Organisations	■ obtain data, software and hardware at minimum cost so available funds can be devoted to organisational objectives ■ disseminate information widely to help meet objectives
Academic sector	■ produce published papers on the basis of research or observation ■ challenge the 'taken for granted' views of others ■ teach knowledge, use of tools and foster understanding of GIS/GI among students
All individuals	■ altruism ■ obtain career, finance or status benefits ■ enhance personal skills, competence and knowledge development

Table 3.4   continued

'Player'	Alternative long term agendas
Central Government organisations	▪ as in short term, subject to change of statute, regulations or policy of government
Local Government	▪ as in short term, subject to change of statute, regulations or policy of government
Commercial sector - information trader / publisher	▪ trade profitably and have positive cash flow ▪ minimise costs of getting data from elsewhere ▪ build customer dependency and minimise competition
Commercial sector conglomerate	▪ corporate aims and goals generally more important in long term and progress towards them is monitored centrally
Non- Governmental / Not for Profit Organisations	▪ obtain data, etc at minimum cost so available funds can be devoted to organisational objectives ▪ disseminate information widely to help meet objectives ▪ influence government policies wherever possible
Academic sector	▪ build reputation for expertise in research or teaching to influence grant allocations and enhance 'brand image' ▪ enhance human knowledge ▪ create more highly trained students for work elsewhere
All individuals	▪ altruism ▪ obtain career, finance or status benefits ▪ enhance personal skills, competence and knowledge

Source: Rhind (1998: 208)

This summary suggests that the perceptions of various actors from the public and the private sector may differ substantially in terms of finance and the dissemination of information, both in the short and in the long term.

Under 'Research Initiative 9: Institutions Sharing Geographic Information' of the National Center for Geographic Information and Analysis (NCGIA) in the United States, a focus group considered the organisational aspects of sharing (Onsrud and Rushton, 1992; Batty, 1992; Onsrud and Rushton, 1996). The outcome is a two-dimensional matrix of organisational issues and organisational forms to measure the degree and impact of spatial data sharing (Onsrud and Rushton, 1992). Research in this case is directed to investigating the complexities of dynamic interactions at work in sharing activities.

Similarly, Nedovic-Budic and Pinto (1999a,b) have proposed a conceptual framework, based on Kevany (1995) and a literature review, consisting of four general theoretical constructs, namely context, motivation, co-ordination mechanisms and outcomes. In their subsequent research, they have used case study and survey methods based on this framework. In particular, the empirical research on the nature of interorganisational sharing arrangements, has focused on sharing initiatives that are ongoing or under development (Nedovic-Budic and Pinto, 2000) and sharing 'clusters' of organisations (Nedovic-Budic *et al.*, 2001). However, in both cases, Research Initiative 9 and the Nedovic-Budic and Pinto framework, the focus on actual incidents of sharing (albeit successful and unsuccessful sharing activities) limits the scope of the research to an investigation of the views of the 'sharers'. 'Non-sharers' are excluded and hence important insights into *why* individuals within organisations may not be willing to share cannot be captured.

The above discussion highlights two essential issues for this research. First, the determinants of whether and why organisations may be willing to engage in spatial data sharing with other organisations need to be established empirically, they should not be assumed a priori. Second, the scope of empirical research should encompass the whole community of potential sharers involved in GIS. If the fundamental factors underlying a disposition to engage in spatial data sharing can be identified, this can provide a basis for the formulation of appropriate mechanisms to encourage spatial data sharing behaviour.

Choice of Theoretical Approach

The foregoing discussion highlights the fact that it would be fruitful to explore the issue of spatial data sharing from a perspective that focuses on the various positions that individuals in different organisations take towards spatial data sharing and on why they take these positions. Some theoretical constructs are needed to understand these individuals' and organisations' actions and their behaviour. Therefore, this section justifies the selection of the theoretical approach by considering alternative decision making models, and more specifically, alternative models of attitude-behaviour relations within social psychology. Furthermore, the appropriate level of analysis is established.

Decision Making Research

Research into factors that influence human judgement and decision making provides a basis for deriving the necessary theoretical constructs. This is an interdisciplinary field drawing on contributions from economics, political science, organisation and management studies, and social psychology. The starting point for much decision making research is rational choice theory (Medin and Bazerman, 1999; Abelson and Levi, 1985) and much of the research has focused on the comparison of actual decision making with principles of rationality in decision making (Dawes, 1998). Rational choice theory assumes that preferences and constraints affect behaviour and that individuals optimise in some way (Opp, 1999). The narrow assumptions about a fully informed, rational decision maker have given way to the realisation that cognitive as well as non-cognitive factors influence the decision making process (Keren, 1996; Mellers *et al.*, 1998) and that decision makers are not necessarily fully informed; that perceived, subjective and not just objective, tangible constraints may be relevant; and that constraints and preferences taken together rather than individual constraints on their own may explain behaviour (Opp, 1999).

Distinguishing between the many different approaches to decision making are structural and process models (Abelson and Levi, 1985).[16] Structural models are concerned with *what* decision makers choose while process models analyse the intervening steps in cognitive processes. A further, though less clear, distinction can be drawn between normative and descriptive models (Keren, 1996; Stevenson *et al.,* 1990). The former model considers how decision makers *should* make decisions and the latter model how they actually *do* make decisions. In the light of the overall goal of this research to examine the nature of the conditions under which the different actors involved with spatial data are willing to engage in spatial data sharing, a structural, descriptive model was considered most appropriate. This enables an investigation of *what* decisions individuals within organisations actually *do* take with respect to spatial data sharing.

Fitting this requirement are prospect theory and expected value models. Prospect theory developed by Kahneman and Tversky (1979) takes into account contextual factors but it assumes the existence of an idealised individual (Abelson and Levi, 1985) and suffers from ambiguity (Yates, 1990). Expectancy value models make no assumptions about rationality and instead rely on the internal consistency between the constructs included in the models (Ajzen, 1996). Expectancy value models are not restricted to cognitive elements and allow for the inclusion of non-cognitive factors such as emotions and desires (Ajzen, 1996).

Much behavioural decision research relies on revealed, observed preferences where probabilities and values have to be inferred from people's judgements (Medin and Bazerman, 1999). The contribution of social psychology to this field is the development of direct measurements of perceived preferences and constraints (Jones, 1985; Ajzen, 1996; Opp, 1999) which have been developed since the 1930s (Hogg and Vaughan, 1995).[17] The expectancy value models of attitudes used by social psychologists are employed in this study in order to provide a basis for analysing the considerations that may underlie real life decisions to share spatial data across organisational boundaries.

Attitude-Behaviour Models

Research within the discipline of social psychology deals with decision making in the general context of predicting and explaining behaviour and the research on attitudes is a central consideration within this body of work. Beliefs are understood as providing the subjective basis for individuals' decisions. The source of beliefs may be logical processes as well as emotions or desires (Ajzen, 1996). Attitudes are assumed to reflect the beliefs that the decision maker holds (Eagly and

---

[16] This parallels the distinction between variance and process model in more general literature on organisational behaviour (for example, Mohr, 1982).

[17] For a discussion of different measurement techniques, see Hogg and Vaughan (1995).

Chaiken, 1998) and the relationship between attitude and actual behaviour is taken to be mediated by the intention to act (Fishbein and Ajzen, 1975).

Several models of the attitude-behaviour relationship have been developed that examine the beliefs that influence attitude formation. The framework for examining this relationship that has received widespread attention in the literature is that which has given rise to the use to expectancy value models that characterise the relationship between beliefs and attitudes. Expected values are said to be made up of subjective probabilities of outcomes and subjective evaluations (positive or negative) of outcomes of a behaviour. The most salient, the Theory of Reasoned Action (TRA) proposed by Fishbein and Ajzen (1975), is most appropriate when behaviour is subject to volitional control and, apart from attitudes, it contains a social norm component to capture situational constraints that may influence decisions. Ajzen (1991) has extended the TRA to account for planned, more complex behaviours to formulate the TPB. Neither, the TRA nor the TPB, assume that decision makers necessarily engage in elaborate cognitive processes prior to taking action (Fishbein and Ajzen, 1980; Ajzen, 1991); instead, an individual has formed an attitude by thinking about the consequences of a given behaviour and those attitudes or intentions can be retrieved and acted upon at a later time. The TRA and the TPB are not restricted to explaining the determinants of behaviour in a specific behavioural domain, they can be widely applied.

An alternative model is Eagly and Chaiken's (1993) composite model of attitude-behaviour relations. The composite model differs from the TRA and the TPB by distinguishing between attitudes towards a behaviour and attitudes towards the *target* of the behaviour. However, attitudes towards a behaviour, as included in the TPB and the TRA, have been found to be particularly effective in predicting actual behaviour (Eagly and Chaiken, 1998; Ajzen, 1991). The composite model does not include control factors as an explicit term: these are assumed to affect the link between attitude and intention. Since the TPB includes variables that capture the individual's perceived behavioural control to allow for behaviour that requires skills, resources and the co-operation of other people, it provides an appropriate framework within which to examine spatial data sharing behaviour across organisational boundaries. Moreover, among the contending theories, the TPB framework is the most widely applied and tested on a wide variety of non-volitional behaviours (Stahlberg and Frey, 1996).[18]

Modifications to the TPB have been discussed, such as the inclusion of personal norm and perceived moral obligation, self-identity, and past behaviour variables, to improve the predictions of intention and behaviour (Sabini, 1995; Eagly and Chaiken, 1993). It is argued that these additions may be drawn upon and included in the TPB model depending on the specific behaviour under investigation (Eagly and Chaiken, 1998).

---

[18] The current list of empirical research papers applying the TPB exceeds 240 articles (http://www-unix.oit.umass.edu/~aizen).

Level of Analysis: Macro and Micro Considerations

Pfeffer (1985) has suggested that when trying to understand organisational behaviour and decisions, a focus on the individual as the unit of analysis often leads to a neglect of normative contexts and technological aspects. Jones (1985) has stressed:

> The individual must be seen as the intersection point of a variety of pressures: immediate situational demands, conflicting social expectations, and internalized beliefs and values. (Jones, 1985: 53)

Using the TPB as an organising framework, the influence of these aspects may be explicitly addressed by the inclusion of social norm and control components within the model. Moreover, the suitability of psychological theories to understanding organisational behaviour has been discussed by Staw (1991) and more specifically for the TPB by Elliot *et al.* (1995). Staw (1991) has suggested that psychological theories that typically examine behaviour at the micro level can also be used to understand action at the macro level, i.e. organisational behaviour.

> ...because it is possible to identify key actors in important organizational decisions, psychological research can be applied to these individuals in order to explain organizational actions. (Staw, 1991: 812)

According to Staw, the most fruitful approach is not to examine in detail all individual behaviour within an organisation, but rather to study the key organisational decision makers. Bacharach *et al.* (1995) argued that because actors making decisions are accountable for their decision, they seek decision criteria that can be used to justify those decisions. These key decision makers may be assumed to have an accurate understanding of their organisation's position towards spatial data sharing and their perceptions may be the best indicator of their organisation's behaviour (Elliot *et al.*, 1995). In this way, the focal situation can be expanded to a larger arena (Markus and Robey, 1988), in this case, to the scope of a national spatial data infrastructure. This position with respect to the representation of individuals as indicators of likely organisational behaviour provides the basis for the research strategy adopted in this study. Rather than limiting the research to an in-depth study of a few organisations, a range of individuals and organisations has been included with the goal of examining their dispositions towards spatial data sharing.

The Theory of Planned Behaviour as an Organising Framework

This section introduces the principles of the TPB that provided the organising framework for this study. As stated above, the TPB (Ajzen, 1985; 1986; 1988; 1991) is an extension of the Theory of Reasoned Action (Fishbein and Ajzen, 1975; Ajzen and Fishbein, 1980) and is most appropriate for the investigation of behaviours that are not under volitional control, i.e. performance of the behaviour is not only reliant on the intention to carry out the behaviour but also on opportunities and resources (Ajzen, 1988). Spatial data sharing among different organisations cannot be assumed to be under volitional control because certain skills, resources or opportunities may play a role in determining whether they can engage in spatial data sharing across organisational boundaries.

Basic Principles of the Theory of Planned Behaviour

The TPB consists of five distinct components; a particular behaviour under consideration, the intention to act, and three determinants of intention (attitude, subjective norm and perceived behavioural control). A distinction is made between the decision to engage in a particular behaviour, conceptualised as behavioural intention, and actual performed behaviour.

Ajzen (1988) argued that:

> Intentions are assumed to capture the motivational factors that have an impact on a behavior; they are indications of how hard people are willing to try, of how much of an effort they are planning to exert, in order to perform the behavior. These intentions remain behavioral dispositions until, at the appropriate time and opportunity, an attempt is made to translate the intention into action. (Ajzen, 1988: 113)

Because of the distinction between intention to act and actual behaviour, the intention construct provides the basis for investigating the disposition of organisations towards spatial data sharing. Hence, it is possible to measure the intention or the willingness to engage in spatial data sharing which is, conceptually, expected to be closely linked to actual sharing behaviour.

The first determinant of intention, attitude towards the behaviour, is a person's positive or negative evaluation of performing the behaviour. The subjective norm captures the individual's perception of social pressure to engage or not to engage in the behaviour. Finally, perceived behavioural control consists of the perceived availability of required opportunities and resources to perform the behaviour.

These components have been tested empirically to predict intentions and behaviour (e.g. Ajzen, 1985, 1988, 1991; Ajzen and Madden, 1986; Ajzen and Driver, 1992). The relative weight of the attitudinal, normative and control factors is expected to vary according to the behaviour under investigation. As the TPB aims to *explain*, rather than merely *predict*, a behaviour, it can be used to trace the determining antecedents of attitude, subjective norm and perceived behavioural control. The underlying foundation of beliefs about the behaviour is considered to provide the basis for perception.

> It is at the level of beliefs that we can learn about the unique factors that induce one person to engage in the behavior of interest and to prompt another to follow a different course of action. (Ajzen, 1991: 206-207)

The TPB, like the underlying Theory of Reasoned Action, focuses on subjective perceptions of individuals rather than on objective observations to explain behaviour.

Basic to this approach is the view that people use the information available to them in a reasonable manner to arrive at their decisions. This is not to say that their behavior will always be reasonable or appropriate from an objective point of view. People's information is often incomplete and at times also incorrect. But we would argue that a person's behavior follows quite logically and systematically from whatever information he happens to have available. (Ajzen and Fishbein, 1980: 244)

Although people are assumed to hold a great number of beliefs about a particular behaviour, only a small number, the so-called *salient* beliefs, are expected to be the predominant determinants of intention and action. According to the expectancy value principle, it is necessary for each belief and its strength (likelihood) to be rated and then combined.

Three types of beliefs can be distinguished; *behavioural* beliefs (influencing attitude), *normative* beliefs (determining social norm), *control* beliefs (underlying perceived behavioural control). Behavioural beliefs consist of the evaluation of specific consequences and outcomes that may result from performing the behaviour in question. Normative beliefs constitute beliefs about the important referents (individuals or groups) and their likely approval or disapproval with regard to engaging in the behaviour. Control beliefs capture the perceived presence or absence of requisite resources and opportunities such as skills and capabilities and the co-operation of other people deemed necessary to perform the behaviour. A behaviour is said to be explained once the determinants of intention and behaviour have been traced to the underlying beliefs. The detailed descriptions developed at the belief level of analysis provide a systematic basis for inferring why individuals embedded within organisations may be willing, or resistant, to engaging in spatial data sharing with other organisations.

The TPB posits a link between perceived behavioural control and actual behaviour. This is because the intention-behaviour relation is conceived to be strong only if the behaviour in question is under volitional control.[19] When this is not the case, a strong intention to perform the behaviour and actual control over the required resources and opportunities for performing the behaviour are considered to influence whether the behaviour will actually occur. Since it is problematic to assess *actual* control, Ajzen included *perceived* control as a proxy for actual control and posits a direct link between perceived behavioural control and behaviour.

To the extent that perceptions of behavioral control correspond reasonably well to actual control, they should provide useful information over and above expressed intentions. (Ajzen 1988: 133)

---

[19] In this case, the use of the TRA (Fishbein and Ajzen; 1975; Ajzen and Fishbein, 1980) would be sufficient.

Finally, a fundamental principle of the TPB that is included to ensure the correspondence of measures is the 'principle of compatibility'. This is based on Fishbein and Ajzen's (1975) work, where all the elements of the model, from beliefs to intention to behaviour, must be assessed at the same level of generality or specificity. This theoretical assumption has to be met in order for the application of the TPB to yield meaningful results.

Additions and Extensions

The TPB has been scrutinised in the literature and several additions have been suggested (see for example, Eagly and Chaiken, 1993,1998; Bagozzi and Kimmel, 1995; Elliot *et al.*, 1995; Giles and Cairns, 1995; Hogg and Vaughan, 1995; Parker *et al.*, 1995; Sabini, 1995; Terry and O'Leary, 1995; Stahlberg and Frey, 1996; Conner and Armitage, 1998; Sutton, 1998) such as the inclusion of self-identity, morality, and past behaviour. Ajzen (1991) has argued that the model is open to further elaboration of additional constructs. Rather than including all possible additional variables in the TPB at once, Conner and Armitage (1998) have suggested that different combinations of these variables may be appropriate depending on the nature of the behaviour under study. Two variables, morality and past behaviour, were included in the model developed in this study.[20]

Morality may be expected to have an influence on the performance of behaviours with a moral dimension (Conner and Armitage, 1998). The influence of perceptions about morality is not likely to be mediated by the basic components of the TPB (Stahlberg and Frey, 1996). Perceptions about morality can be considered either as moral *implications* (i.e. outcomes) (Parker *et al.*, 1995) or moral *obligations* to perform or not to perform a particular behaviour (Sabini, 1995). As moral implications or evaluations of a behaviour, morality would be considered an antecedent of attitude (Bagozzi, 1989; Parker *et al.*, 1995). As moral obligations, it would be an underlying determinant of subjective norm (Sabini, 1995). Alternatively, it could be regarded as a variable that is independent of either attitudes or subjective norms (Conner and Armitage, 1998). Since no consensus has been reached as to where exactly perceptions of morality should be placed conceptually in the model of the TPB, in this research, perceptions about morality have been considered both in the form of social outcomes influencing attitude and as moral norms determining subjective norms. Specific measures of response concerning the moral determinants of attitudes and subjective norms developed by Bagozzi (1986) have been drawn upon in this study (see Chapter 5).

Past behaviour has been discussed as a proxy of habit for highly routinised behaviours (Sabini, 1995; Eagly and Chaiken, 1998). Bagozzi and Kimmel (1995) have argued that as a behaviour is executed repeatedly, it is influenced more by habit and less by intentions. In contrast, Conner and Armitage (1998) maintain that:

---

[20] Self-identity was not included because this implementation of the TPB was aimed at capturing the key decision makers' perceptions of their organisation's behaviour so the relation of the actor's self-identity to spatial data sharing was not of principal interest.

However, frequent performance of a behavior may bring subsequent behaviour under the control of habitual processes, although a behavior does not necessarily become habitual just because it has been performed many times. (Conner and Armitage, 1998: 1436)

Similarly, Ajzen (1991) suggests that '... past behavior is best treated not as a measure of habit but as a reflection of all factors that determine the behavior of interest.' (Ajzen, 1991: 203).

Furthermore, 'habitual behaviors may not be amenable to prediction by models such as the TPB' (Conner and Armitage, 1998: 1438). Measures of past behaviour have been included in the model for this research because organisations may have engaged in spatial data sharing in the past and, in line with Ajzen's (1991) argument, the model should take the influence of their past experience with spatial data sharing into account. The items to assess past experience were adapted from Metselaar (1997) and are discussed further in Chapter 5.

Similar to the discussion on the position of perceptions of morality in the model, the position of past behaviour in relation to the basic components of the TPB is subject to debate. Bagozzi and Kimmel (1995) have investigated the effects of past behaviour which may be mediated by the other components of the TPB or which are assumed to take effect more directly. Ajzen (1991) has argued that the addition of past behaviour as a separate variable should not significantly improve the prediction of later behaviour if all important variables are contained in the set of determinants. He maintains that past behaviour should not be considered a predictor of later behaviour: '... although past behavior may well reflect the impact of factors that influence later behavior, it can usually not be considered a causal factor in its own right' (Ajzen, 1991: 203).

However, past experience is an important source of information about perceived behavioural control such that perceived behavioural control may have a mediating effect of past behaviour on later behaviour (Ajzen and Madden, 1986). In this study, the determinants of perceived behavioural control include an assessment of past experience with spatial data sharing to allow for this mediating effect (see Chapter 5).

Research Questions Addressed within the Organising Framework

As spatial data are indispensable for the use of GIS, the principal aim of this research is to examine the nature of the conditions under which the different actors involved with spatial data are likely to be willing to share the data across organisational boundaries to improve the availability of, and access to, spatial data to use GIS more effectively.

In the light of the theoretical framework adopted for this study, the principal research question proposed above can be examined by reformulating it in operational terms as follows:

• How willing or resistant are key individuals within different organisations to engage in spatial data sharing across organisational boundaries?

- What is the extent of spatial data sharing activities that are already taking place?
- What are the factors that determine the willingness or resistance of key individuals within organisations to engage in spatial data sharing?

The constructs of the TPB provide a framework within which two of these questions can be addressed. In particular, the intention construct can be used to address the first research question and to assess the willingness of individuals embedded within organisations to share spatial data with other organisations. The determinants of intention (attitude, subjective norm, and perceived behavioural control) and their underlying belief structures provide the constructs necessary to examine the specific determinants of this apparent willingness, thereby addressing the third research question. The focus on the behaviour under investigation provides the scope for addressing the second research question regarding actual spatial data sharing activities of organisations.

This framework provides a systematic way of combining qualitative and quantitative empirical research methods to develop a model of the propensity to engage in spatial data sharing across organisational boundaries. Specific aspects of the research design for this study are considered in the next chapter.

Conclusion

This chapter has provided the theoretical framework for the investigation of spatial data sharing that is an important issue for the success of spatial data infrastructure initiatives that are underway in many countries. In order to investigate whether spatial data sharing is likely to follow from initiatives that are designed to promote standard setting, the establishment of clearinghouses and consistent pricing policies for the national spatial data infrastructures, it has been argued that it is necessary to identify the existing incentives and disincentives to participating in data sharing that may be present within the communities of actors, and their interactions, with respect to the generation and use of spatial data.

The literature on spatial data sharing includes an array of concepts and arguments that aim to explain why people or institutions do or do not share spatial data. In this chapter, the selection of the TPB as an organising framework for investigating some of the underlying facets of spatial data sharing (outcomes, social pressures and expectations, and resources and opportunities) was introduced. This theory provides the basis for formulating a structured model and for selecting appropriate research tools to enable the prediction and explanation of spatial data sharing behaviour within a specific context. The results of this approach also provide a basis for specific guidance as to how policy makers may influence the actual behaviour of spatial data sharing more effectively.

Using South Africa as a case study, the TPB model is applied to elicit the dispositions of the members of the GIS community to engage in spatial data sharing across organisational boundaries. Specifically, it is applied to investigate why they think they may stand to gain or lose as a result of spatial data sharing;

what they think those who are important to them (inside and outside their organisation) want them to do; and what control in terms of skills and capabilities they perceive that their organisations have over whether or not they get involved in spatial data sharing. Once the profile of perception is established, this provides a basis for addressing the issue of spatial data sharing more effectively at the policy making level within the framework of the NSIF in South Africa.

Chapter 4

Research Design and Implementation Methods

Introduction

The previous chapter introduced the organising framework for this research. Several methodological issues require consideration in order to operationalise this framework. A distinction is made between two phases of empirical research; a phase of qualitative interviews to elicit beliefs and a second, quantitative phase consisting of a survey. This chapter provides details of the steps needed to implement the TPB for an analysis of spatial data sharing behaviour.

The first section provides a detailed specification of the research design including, first, the choice of the country, South Africa, as a case study within which the empirical research took place; second, a definition of the behavioural elements that are included in the model; and third, the two stages of the empirical research. The following section presents the methods used to implement these two stages of empirical research. The selection of interviewees and survey respondents, the design of the survey questionnaire, and sampling procedures are discussed. Finally, the last section concludes the chapter with reflections on the methodology.

Research Design for Implementing the Organising Framework

Choice of Country

The empirical research was carried out in a single country. The deciding factor for limiting the research to one country was the need to conduct the research in sufficient depth to reveal the specific beliefs and behaviour of a representative sample of actors and organisations as required by the theoretical framework. Time and resource constraints meant that this could be undertaken only for a single country but it is recognised that future research is required to determine whether the pattern of beliefs and behaviour that is revealed in this study is likely to vary substantially in other national contexts. The research for this study was conducted to analyse in detail the determinants of spatial data sharing behaviour in South Africa. Classified by the United Nations as a 'middle income' developing country (World Bank, 1999), the difficulties of political restructuring in South Africa since the end of the apartheid regime are accompanied by very high levels of inequality in terms of education, income, and land ownership. The principal criteria that influenced the selection of South Africa are as follows.

First, the focus of the study mandated an exploration of the determinants of spatial data sharing in a *developing* country context where resources for generating spatial data can be expected to be relatively limited and therefore there may be stronger incentives to engage in spatial data sharing to reduce costs than in a country where greater resources are available. Second, in order to explore the factors determining spatial data sharing, GIS had to be in use in the selected country. This was the case in South Africa where considerable investments in GIS have been made and they are being used relatively widely in public, private and non-profit organisations. The diffusion of GIS has been accompanied by the appearance of spatial data bottlenecks in terms of limitations on the required quality, availability, and scale of spatial data. Similar to the efforts in many other countries, the NSIF initiative was invoked to try to overcome these bottlenecks by facilitating spatial data sharing (for details see Chapter 2). Third, access to spatial data is restricted in some countries, such as India, where maps are treated as classified documents by the government, and research into GIS is therefore restricted by concerns about national security. The South African government, on the other hand, very openly encourages the use of geographic information and actively pursues the proliferation of geographic information.

Definition of Behavioural Elements

This section presents the definition of the behavioural elements that are required to develop a model of spatial data sharing. The TPB requires the identification of four elements to define a given behaviour: action, target, context and time. Ajzen (1985, 1991) places particular importance on the correspondence of these elements across behaviour, intention, attitude, subjective norm and perceived behavioural control, and also with the underlying beliefs. Therefore, the precise definition of each element provides reference points which can be used in the elicitation of beliefs in the first stage of empirical research and in the subsequent construction of a questionnaire. The questionnaire is used in the second stage of empirical work to generate empirical evidence of the determinants of the willingness to engage in spatial data sharing. The following provides the definition of each element of a behaviour, in this case, spatial data sharing.

Action To clarify the focus of the study, the behaviour under investigation must be specified. The behaviour in this instance is spatial data sharing, as defined in Chapter 3,[21] i.e. spatial data sharing entails making the digital spatial data used in GIS accessible to or from other parties. These exchanges may or may not include barter, financial payment or payment in kind.

Target The definition of the target required consideration of the usage and definition of spatial data, spatial information, geographic data or information and geospatial data or information. These terms are often used interchangeably and a number of different uses can be identified.

Spatial data are geographically referenced features that are described by geographic positions and attributes in an analog and/or computer-readable (digital) form. (Frederick 1995: 362)

The United States Federal Geographic Data Committee (FGDC), in charge of implementing the National Spatial Data Infrastructure (NSDI) in the US, refers to digital geospatial data which it has defined as the digital 'information which identifies the geographic location and characteristics (attributes) of natural and constructed features and boundaries on the earth' (FGDC 1997: 71). A hierarchy of definitions is given by Environment Systems Research Institute, one of the major GIS software suppliers, which defines geographic data as 'The locations and descriptions of geographic features. The composite of spatial data and descriptive data' (ESRI, 1993: G-14). Spatial data, in turn, refers to 'Information about the location and shape of, and relationships among, geographic features, usually stored as co-ordinates and topology' (ESRI, 1993: G-43) and descriptive data are 'Tabular data describing the geographic characteristics of map features. Can include numbers, text, images and CAD drawings about features' (ESRI, 1993: G-9). In the academic literature on GIS, geographic data or information, and spatial data or information are terms that are used interchangeably. According to Burrough and McDonnell (1998), such data and information represent phenomena from the real world in terms of:

- their position with respect to a known co-ordinate system (spatial co-ordinates or geometry),
- their attributes that are unrelated to position (properties),
- and their spatial interrelations with each other which describe how they are linked together (topology).

Similar terms and the term 'spatially referenced information' are used in the NSIF material, incorporating both locational and descriptive components of the data rather than referring to them separately. This lack of clarity and the absence of precise terminology suggested that firstly, there could be some flexibility with

---

[21] The term 'spatial data sharing' is used extensively in the context of the NSIF in South Africa and other national spatial information initiatives. The use of this term throughout the empirical part of the study could therefore be expected to receive appropriate interpretation from the interviewees.

regard to the way terms are used in this study. The term most often used in the NSIF material is 'spatial data' and, hence, this term was adopted here, including the clarifying prefix 'digital'. Secondly, the predominant terms in use tend to include both locational and descriptive components of the data and even topological features and are somewhat more consistent in this logical sense. This definition encompasses a long list of different geographic information types, all of which have a locational dimension.[22] In conclusion, the target of the behaviour in question is digital spatial data, incorporating locational and descriptive elements.

Context The spatial data sharing behaviour under consideration in this study was sharing across organisational boundaries.

Time The time dimensions taken into account were the present (at the time of questionnaire implementation in July-August 1999) and the short term (covering the period of the next month to two years from the present).
Combining these behavioural elements, the following reference point (see Table 4.1) was created as a basis for developing measures of the constructs.

Table 4.1   Research specification

Action	Making spatial data accessible to or from other parties
Target	Digital spatial data (incorporating locational and descriptive elements)
Context	Spatial data sharing across organisational boundaries
Time	Present and short term (in the next month to two years)

Research Design

The research design developed for the application of the TPB to investigate the determinants of spatial data sharing behaviour involved a combination of qualitative and quantitative empirical research methods (see Figure 4.1). The choice of these methods was considered appropriate in the light of the principles that underpin the TPB.

- In the *qualitative* empirical stage, items to provide a proxy measure of each theoretical construct were identified in semi-structured interviews with a sample of individuals in South Africa to ensure that the identified beliefs would have a concordance with the spatial data context under investigation. The analysis of this interview material was complemented by the insights

---

[22] Although the FGDC and the NSIF were important sources for the definition of spatial data sharing, it is necessary to point out that this study was not intended to be an evaluation or feasibility study of these initiatives. Rather, it was intended to further the understanding of the determinants of spatial data sharing behaviour - the behaviour which these initiatives in their various forms hinge upon and aim to foster.

derived from a review of several relevant bodies of literature and this material provided the basis for the development of a model of spatial data sharing.

• The model of spatial data sharing was then operationalised by developing items to be included in a questionnaire. The next empirical stage of the research consisted of the implementation of the questionnaire instrument in South Africa. The survey data was analysed using quantitative methods to investigate the factors that seem to influence the willingness of key individuals within their organisations to share spatial data across organisational boundaries.

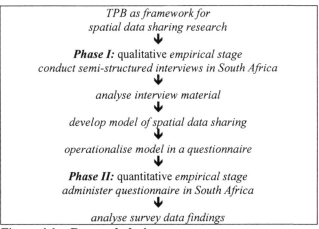

Figure 4.1   Research design

Operationalising the Organising Framework

Selection of Interviewees and Survey Respondents

As outlined in the previous section, this study consisted of two stages of empirical research. A precondition for the selection of the organisations for both stages was the existence of an operational GIS (Calkins and Weatherbe, 1995). This criterion was considered appropriate to ensure that all of the respondents were actively involved in spatial data applications at the time of the study rather than considering their future engagement as a user, supplier or value-adder.

The selection of the interviewees in each organisation was made on the basis of their professional position and experience. Selected interviewees were key individuals who made decisions about access to spatial data to or from other parties. For the survey, it was deemed desirable to include more than one person in each organisation who was knowledgeable about the history and environment of the organisation to identify and limit any bias that might be introduced as a result of the discrepancy between the views of a single person and the organisation as a whole (Kevany, 1995).

Implementation of Phase I

Phase I of the empirical research was carried out over a two week period (4 - 17 March 1999) in two provinces (Gauteng and the Western Cape) in South Africa. Most of the interviews were set up in advance using contacts that had been established via email, while a few additional interviews were arranged during the fieldwork. Twenty semi-structured interviews were conducted with a view to eliciting the different types of beliefs (behavioural, normative and control beliefs). The interview questions were based on and adapted from the questions suggested by Ajzen and Fishbein (1980) and Ajzen (1985, 1991) for this elicitation procedure. A list of the questions asked is presented in Appendix A. The responses were recorded in written format and subsequently digitised for further processing.

As identified by Abbott (1996) and confirmed during the interviews in Phase I, there are distinct groups within the GIS community in South Africa that stem from the wide range of possible applications of GIS and the actual spread of GIS in South Africa. The different sectors included in the GIS community are presented in Table 4.2, together with the distribution of the qualitative interviews in Phase I.

Table 4.2   Phase I - Distribution of interviews by sector

Sector	no. of interviews
Local authorities	0
Provincial government	3
National government	6
Para-statal organisations	4
Academic / research institutes	3
GIS industry	3
Private sector	0
Non-governmental, not-for-profit	2
Total	**21**

Although not every sector in the GIS community in South Africa could be represented in this stage of the research, the range of interviewees can still be considered sufficient to ensure concordance of the identified beliefs with the spatial data context under investigation because six out of eight sectors were included.

Analysis of Interview Material from Phase I and Questionnaire Design

The qualitative interview material from the first phase of empirical research was analysed to develop a model of spatial data sharing, complemented by secondary material and relevant theoretical bodies of literature. Specifically, the procedure consisted of the following steps.

First, in the process of analysing the interview material, the information was extracted by collating the responses of all the interviewees to each question. In this way, the most common behavioural, normative and control beliefs could be identified. According to the nature of the beliefs - depending on whether they were addressing outcomes of spatial data sharing (i.e. behavioural beliefs), referents for spatial data sharing (i.e. normative beliefs), or skills, resources and opportunities for spatial data sharing (i.e. control beliefs) - the beliefs were grouped into *domains*. The resultant domains were assigned to the corresponding main components of the TPB (attitude, social pressure and perceived behavioural control).

Second, where beliefs to be included in the model could not immediately be operationalised, secondary material and relevant theoretical bodies of literature were consulted. The secondary material consisted of minutes from NSIF workshops where the most pressing issues had been voiced by a variety of stakeholders (Clarke *et al.*, 1998; and personal notes from a NSIF workshop attended by the author). Material relating to national spatial data infrastructures in other countries (MSC, 1993, 1994; FGDC, 1997; Burrough and Masser, 1998; Masser, 1999; Rhind, 1999) was used to complement and support these insights and to support the choice of the most salient beliefs to be included in the model. Once collected, the constructs were further informed by theoretical insights from relevant theoretical literature and a detailed model of the determinants of the willingness of organisations to engage in spatial data sharing was compiled. This material is discussed further in Chapter 5. The extent to which the interview material was combined with secondary material and relevant theoretical literature varied for different constructs. This means that the sections in Chapter 5 detailing the components of the complete model differ substantially in length.

Third, a questionnaire was designed to obtain measures of the constructs contained in the TPB in relation to perceptions of spatial data sharing across organisational boundaries. The model of the willingness to engage in spatial data sharing provided the basis for constructing the questionnaire instrument. The components of the model were operationalised to provide proxy indicators and questions that were included in the questionnaire (see Appendix B). Similar to the approach used for constructing the model, the formulation of the questions for the questionnaire instrument was aided by the secondary information from the NSIF workshops and other national spatial data infrastructure developments around the world. The final questionnaire contained both questions to measure the constructs, and some demographic questions in order to distinguish among the respondents. Unipolar and bipolar scaling was employed in questionnaire item construction depending on the construct under investigation. A pilot test of the questionnaire was conducted during the first week of the survey implementation in Pretoria and Johannesburg, South Africa, to further improve the questions, modify or delete some items, and to obtain an estimate of the time required to complete the questionnaire (45-60 minutes).

The semi-structured interviews in Phase I had revealed that respondents had a tendency to answer the questions in relation to GIS more generally rather than to spatial data sharing specifically, even though the questions explicitly asked about spatial data sharing. This necessitated that the questionnaire be as specific as possible about spatial data sharing. The advantage of the survey work in Phase II was that interviewees could *read* the questions to help them focus on the subject of enquiry. In other areas of application of the TPB, the questionnaire may be more elegant, i.e. less repetitive, because respondents will associate the questions with the behaviour in question. However, for this study it was necessary to be as explicit as possible to ensure that respondents answered the questions posed.

Implementation of Phase II

This section outlines the survey sample design and the procedures used in the implementation of the survey.

Sample design and sampling options The target population for the survey was the GIS community in South Africa. The names and contact details of organisations were drawn from three different databases of GIS-using organisations (as of March 1999) that were made available for this research; the customer database of a spatial data supplier, that of a GIS supplier, and the database resulting from a detailed academic survey of the diffusion of GIS within the public sector in South Africa. Despite the availability of these databases and even after consultation with the NSIF, it was not considered possible to establish the absolute number of organisations in each of the sectors (or that of the whole GIS community for that matter). However, it was possible to establish that, apart from a small number of GIS-using non-governmental organisations (NGOs), the remainder of the GIS community was divided approximately equally among the other sectors. The distribution of the survey respondents in Phase II is shown in Table 4.3. In the time available for conducting the survey, it was considered feasible to aim for a minimum of fifteen questionnaires per sector (except the NGO sector) so as to attain a sample of at least one hundred respondents.[23] Column two (total number of questionnaires) of Table 4.3 shows that this minimum target was exceeded for all sectors.[24]

The total number of respondents included in the final analysis was 112 from 73 different organisations. The distribution of these organisations by sector is presented in Table 4.4.

---

[23] There are no clear guidelines as to how large a sample should be (Bryman and Cramer, 1999); generally, a sample of less than 60 is considered small (Allison, 1999). Owing to the considerable number of variables arising from the model of spatial data sharing for this study, it was considered appropriate to aim for a sample of at least one hundred survey respondents.

[24] There was no minimum target for the number of questionnaires to be collected in the NGO sector owing to the small number of organisations in this group within the GIS community. Every effort was made to include as many GIS-using organisations in the survey as possible.

Table 4.3   Phase II - Distribution of survey responses by sector

Sector	Total no. of questionnaires collected	Pilot questionnaires	Incomplete questionnaires	Unreliable questionnaires	Total no. for analysis
National government	19	3	0	1	15
Provincial government	18	3	0	3	12
Local authorities	19	0	1	0	18
Para-statal organisations	25	2	1	2	20
Academic / research institutions	19	0	2	1	16
GIS industry	16	0	0	3	13
Private sector	21	2	2	4	13
NGOs	6	1	0	1	4
Other	2	0	1	0	1
Total	**145**	**11**	**7**	**15**	**112**

Note: The responses were categorised according to the following criteria:
1. The questionnaires categorised as *incomplete* had more than 10% missing responses.
2. The categorisation of questionnaires as *unreliable* was based on the assumption that reliably filled in questionnaires would demonstrate correspondence between the belief and the domain level questions. This method involved summing the scores of beliefs in each domain and relating these sums to the domain level score. Questionnaires with discrepancies among beliefs and corresponding domain for 5 or more domains were considered unreliable and excluded from the analysis.

Table 4.4   Phase II - Distribution of organisations by sector

Sector	Total no. of questionnaires for analysis	No. of organisations
National government	15	9
Provincial government	12	7
Local authorities	18	11
Para-statal organisations	20	12
Academic / research institutes	16	11
GIS industry	13	11
Private sector	13	8
Non-governmental, not-for-profit	4	3
Other	1	1
Total	**112**	**73**

A sample that is unrepresentative of the larger population from which it is drawn limits the ability to generalise the findings to the population (Bryman, 1989; Bryman and Cramer, 1999). In the case of this study, the sectoral composition of the GIS community constitutes the descriptive characteristic that had to be represented in the sample in order for the results of the survey to be generalisable to the population, i.e. the GIS community at large within South Africa. A completely random sample was not deemed appropriate due to the uneven sectoral fragmentation in the GIS community which would have meant that marginal groups, such as NGOs, might not have been adequately represented. Hence, a sectoral stratification of the South African GIS community combined with random sampling within the sectors was employed which ensured the representation of the specific groups that exist in the GIS community at large.

A substantial proportion of the survey responses was obtained in face-to-face interviews over a two month period during July and August 1999. In addition, to ensure a satisfactory sample size, a complementary accidental sampling technique was employed at a suitable conference over a 3-day period (Earth Data Information Systems Conference, 12-14 July, CSIR, Pretoria). The combination of this two-tiered approach produced a sufficient sample size while maintaining a reliable representation of the GIS community. Since the conference was at the start of the survey implementation, subsequent selections for the face-to-face interviews could be made on the basis of the stratification criteria presented in Table 4.3, so as to maintain a representative composition of the sample.

Survey implementation - main approach   The stratification criterion was operationalised using the following method. Having considered the composition of the GIS community in South Africa, distinct geographical areas within South Africa were selected for the implementation of the survey. This kind of clustering technique adds certain geographical boundaries to the scope of the sample and is an acceptable technique to reduce interviewing costs and travel expenses (Bryman and Cramer, 1999). As South Africa is divided into nine provinces, a choice of three provinces was considered appropriate both as sub-units for the study and as a reasonable physical distance to be covered.[25] In addition, this choice also ensured that a sufficient number of provincial government departments could be included.

The three particular provinces, *Gauteng, the Western Cape* and *KwaZulu-Natal*, were chosen because some of the GIS community sectors are very location-bound and not evenly distributed throughout South Africa. This is very much so for the national government departments so that their location influenced the choice of the provinces for the study. Specifically, since most national government departments are located in Gauteng and the Western Cape, these two provinces were chosen. Similarly, these provinces, like KwaZulu Natal, also host some of the non-governmental, not-for-profit organisations. Random samples according to the sectoral stratification criteria were drawn from the contact databases for these provinces. Relevant organisations were contacted by telephone and interviews to administer the questionnaire were set up with the key individuals within each

---

[25] The provinces in South Africa are: Eastern Cape, Free State, Gauteng, KwaZulu-Natal, Mpumalanga, North-West, Northern Cape, Northern Province, Western Cape.

organisation. This approach was more successful than sending a letter (via post or email) to explain the research and to ask for participation, followed up by a telephone call to make arrangements for an interview. The overall response rate attained (including the questionnaires distributed at the conference, discussed in the next section) was very satisfactory at 76.3% (see Table 4.5).

Table 4.5   Phase II - Survey response rate

Sector	Number of questionnaires distributed	Number of questionnaires collected	Response rate %
National government	26	19	73.1
Provincial government	24	18	75.0
Local authorities	25	19	76.0
Para-statal organisations	28	25	89.3
Academic/research institutes	26	19	73.1
GIS industry	24	16	66.6
Private sector	27	21	77.7
Non-governmental, not-for-profit	8	6	75.0
Other	2	2	100.0
Total	**190**	**145**	**76.3**

Survey implementation - complementary approach In addition to the face-to-face interviews, an accidental sampling approach was used since the timing of the fieldwork coincided with a relevant conference in South Africa.[26] This conference gathered people from all over South Africa engaged in the application of spatial data techniques and was therefore regarded as being highly appropriate for this approach. A poster stand was set up to draw attention to the survey and to position the survey within the context of this study. In order to target appropriate respondents, interested conference attendees were asked whether their organisation was based in South Africa, whether they currently had a functioning GIS implemented and whether they were one of the key individuals in their organisation with regard to decisions on access to spatial data to or from other parties. Those attendees who agreed to complete a questionnaire were asked to provide their contact details so that any questionnaires that had not been returned by the end of the conference could be followed up and collected at a later stage.

---

[26] Earth Data Information Systems Conference (EDIS), at CSIR, Pretoria, 12-14 July 1999.

Preparation of Survey Data from Phase II for Analysis

The procedure for cleaning the data collected in Phase II using the questionnaire instrument was as follows. Missing values were coded in the data set and frequency statistics (see Chapter 6) were used to identify whether any variables had high levels of missing values. If this had been the case, it would have been necessary to exclude these variables from the subsequent analysis. As the level of missing values was very low (no more than 3% for any variable), all the variables could be included in the analysis. Prior to conducting the analysis, the missing values were replaced with the mean score of the respective variable. Finally, depending on the direction of the scales (i.e., negative to positive versus positive to negative connotation), the scores for some variables had to be reversed so that the direction of the scales was uniform for all of the variables in the data set.

Hypothesis Construction and Model Validation

The test of validity of the model of spatial data sharing constituted a crucial step in this study since the aim was to systematise the determinants of organisations' spatial data sharing behaviour. Only when the model had been demonstrated as being valid could it be relied upon to further the understanding of the incentives and disincentives for key individuals within organisations to engage in spatial data sharing across organisational boundaries. The validation of the model was necessary to demonstrate that the model developed in this study could be used to address the research questions.

Construct validation consists of deducing and testing hypotheses from a relevant theory. The construction of hypotheses provides a way to relate the variables in the study according to the proposed structure of the organising framework, the TPB. (These hypotheses are presented in Chapter 5). Specific tests can then be used to establish that the relationships between the variables are all in the expected direction. While the use of hypothesis construction and validation implies that much is assumed about the mathematical relationships of the variables, the strength of this approach is that it systematises the validation of the model. The verification of the model of the willingness to share spatial data against the empirical data entailed testing the reliability of the measures used in the questionnaire instrument, their conceptual validity as well as construct validation. These tests are discussed in detail in Appendix E.

Conclusion

The reason for relying on secondary material and bodies of literature in addition to the elicited beliefs in Phase I to inform the model is related to the complexity of the behaviour under investigation. In many of the areas where the TPB has been applied (e.g., Ajzen and Madden, 1986; Ajzen, 1991; Terry and O'Leary, 1995), relatively straightforward behavioural options were considered and this enabled the elicitation of salient beliefs during an initial investigation. In the present study, however, the behaviour under examination is more complex, not least because of the technological aspects involved, entailing, for example, consideration of interoperable standards, formats and data quality. Hence, decomposing the underlying beliefs into domains guided by theoretical insights drawn from the literature was coupled with insights gained from personal interviews during the first stage of empirical work. Interviews alone would not have been sufficient to provide information about the range of beliefs or to provide a basis for translating them into measures and questionnaire items. Moreover, the sole reliance on belief elicitation has been called into question with regard to the completeness of covering relevant beliefs (Conner and Armitage, 1998) and consultation of the literature has been found to be useful in the application of the TPB in other contexts (Taylor and Todd, 1995). Therefore, the complementary approach was taken in this research.

As indicated above, the survey was implemented using face-to-face interviews and administering questionnaires during a conference in South Africa. There were several reasons for relying on these methods. First, owing to the complexity of the model of the perceived willingness to share spatial data, the questionnaire instrument was lengthy. Therefore, it was not expected that an acceptable response rate would be attained by merely posting the questionnaire to potential respondents. Second, the research sampling procedure targeted specific individuals within organisations, i.e. the respondents had to be key individuals who made decisions about access to spatial data to or from other parties. In this respect, the use of face-to-face interviews and the conference for the distribution of the questionnaire provided a screening mechanism for ensuring the 'right' person in each organisation was completing the questionnaire. Finally, the personal presence of the interviewer while respondents completed the questionnaire ensured that they were returned promptly and that the respondents could complete any pages they had missed inadvertently.

The next chapter presents the results of the survey and provides the first insights into the influences of the willingness of key individuals within organisations to engage in spatial data sharing across organisational boundaries.

Chapter 5

Model of the Willingness to Share Spatial Data

Introduction

This chapter presents the construction of the model of the potential willingness of individuals within an organisation to share spatial data. The behavioural approach serving as an organising framework was introduced in Chapter 3. This chapter draws systematically on the qualitative interview material from Phase I, complemented by relevant literature, to address the central research question: What factors are influencing the willingness of key individuals within organisations to share spatial data across organisational boundaries?

The model is developed by extending the TPB methodology. The TPB suggests the eliciting of beliefs in interviews (Ajzen, 1991; 1988; 1985). While that approach was followed by this research, the consultation of relevant literatures to complement, support and decompose the identified items constitutes an extension of the TPB methodology. For a discussion of the methodological aspects, see Chapter 4. Although the initial interview phase elicited most of the beliefs, additional beliefs were derived from the spatial data sharing literature. Not all the beliefs were immediately operationalisable. In these instances, a further search of the literature was undertaken to decompose the beliefs in.question and to develop proxy measures for these beliefs. Hence, as already explained in Chapter 4, the following sections vary in length presenting the procedure that was used to operationalise the beliefs.

According to the research specification for this study that was presented in Chapter 4, the main components of the TPB (presented in Chapter 3) can be adapted to understanding spatial data sharing behaviour as presented in Figure 5.1.

In Figure 5.1, the intention construct refers to the *willingness* to engage in spatial data sharing activities across organisational boundaries; 'attitude' refers to the *attitude* towards spatial data sharing; 'subjective norm' is referred to as the *social pressure* to engage, or not to engage, in spatial data sharing; and 'perceived behavioural control' is referred to as the *perceived control* over spatial data sharing activities of key individuals within organisations.

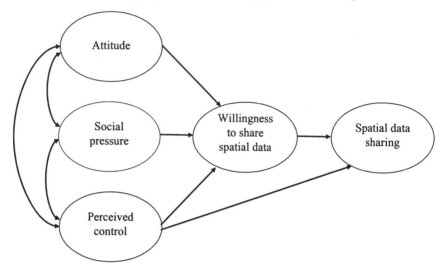

Figure 5.1   Basic model of spatial data sharing
Source: based on Ajzen (1991)

Table 5.1 presents an example of the three 'direct' measures of the willingness of organisations to engage in spatial data sharing that were included.

Table 5.1   Sample questionnaire item: direct measure of intention

In our organisation, we definitely intend to get involved in spatial data sharing. 1. at the moment							
-3	-2	-1	0	1	2	3	☐
strongly disagree	quite	slightly	uncertain	slightly	quite	strongly agree	no opinion /not relevant
2. in the next month - 2 years							
-3	-2	-1	0	1	2	3	☐
strongly disagree	quite	slightly	uncertain	slightly	quite	strongly agree	no opinion /not relevant

The chapter considers in turn each of the components and the underlying beliefs that may influence the willingness of individuals within their organisation to share spatial data. The first section introduces the domains underlying 'attitude', the next section introduces those for 'social pressure', and the third section those for 'perceived control'. Section four presents the complete model of the likely willingness to engage in spatial data sharing. Section five presents a number of hypotheses with respect to the relationships between the constructs that are employed, and the explanatory power of the model. The final section introduces the measures that are used for assessing past spatial data sharing behaviour.

Attitude towards Sharing Spatial Data across Organisational Boundaries

The 'attitude' component arises from evaluations of perceived outcomes of a specific behaviour (Ajzen, 1991). For the case of sharing spatial data, the foundation of beliefs underlying the 'attitude' component was considered to consist of four domains: resources, organisational activities, the strategic position of an organisation, and social outcomes. Each domain under 'attitude' is composed of a set of behavioural beliefs which is introduced. Behavioural beliefs consist of the evaluation of specific consequences and outcomes that may result from sharing spatial data across organisational boundaries. For each behavioural belief, the outcome has to be evaluated as positive or negative and likely or unlikely.

Attitude Domain 1: Resources

The first domain to be considered within 'attitude' is the resource domain. It refers to beliefs about the positive or negative implications of spatial data sharing for an organisation's resources such as time, expense, staff, and administrative effort. The specific beliefs are listed in Table 5.2.

Table 5.2   Behavioural beliefs about resource outcomes

Belief
1.   Time
2.   Expense
3.   Staff
4.   Administrative efforts
5.   Data storage requirements
6.   Introduction of standards
7.   Quantity of spatial data
8.   Quality of spatial data

The evaluation of each belief refers to the organisation's perception of whether each resource is enhanced or reduced as a result of sharing spatial data with other organisations. This evaluation can be assumed to generate a positive or negative attitude towards sharing spatial data. The content of each belief is briefly discussed below.

The beliefs about the implications of sharing spatial data for an organisation's resources are based on the consideration of two aspects. On the one hand, the resources can be seen as those required to *generate* spatial data within an organisation. On the other hand, they are contrasted against those resources required for *retrieving* such information from external sources (Pinto and Onsrud, 1995).

First, with regard to *time,* the amount of time that is saved by sharing spatial data that otherwise would be needed to capture spatial data in-house needs to be evaluated against the time required to locate spatial data outside the organisation, for example through establishing personal contacts or locating spatial data clearinghouses on the Internet.

Second, the financial aspects associated with sharing spatial data can present incentives or obstacles. The *expense* involved in making spatial data available to, or receiving them from, other organisations can include staff time for selecting, downloading, and handling data, computer time, consumable supplies, prorated hardware and software costs, and the staff time involved in assisting users once they have the data (Taupier, 1995). Alternatively, it has been suggested that 'Effective use of spatial data sharing can mean that data are collected in the most logical and cost-effective way' (Calkins and Weatherbe, 1995: 66).

Sharing is argued to reduce the cost of spatial data collection (Alfelor, 1995) and offers a chance to split the cost of common spatial data (Dueker and Vrana, 1995). It is also expected to avoid the downstreaming of costs by reducing the need for elaborate and complex conversions among the databases built and maintained by different organisations (Azad and Wiggins, 1995). Economies of scale can be realised, for example, by undertaking joint conversions of analogue paper-based maps to many layers of digital information that would be more expensive if undertaken separately (Azad and Wiggins, 1995).

Third, specialised *staff* and expertise are necessary for collecting and maintaining spatial data (Onsrud and Rushton, 1995). While sharing may eliminate duplicated or redundant data gathering (Calkins and Weatherbe, 1995; Frederick, 1995), trained staff may still be required for exchanging and integrating spatial data.

Fourth, the *administrative effort* required for sharing can involve strategic policy development, contracts and agreements, and copyright negotiations. Onsrud (1995) has argued that:

> The need to enter into contracts negates much of the willingness and practical ability to share and thus eliminates the benefits to be gained by sharing. (Onsrud, 1995: 301)

This would arise from the administrative burden from developing written contractual relationships and establishing the formal process for dealing with user requests involving lawyers and consultants (Onsrud, 1995). However, it also may be the case that administrative burdens that are incurred through spatial data capture may be reduced by sharing spatial data with other organisations.

Fifth, *data storage* can become an issue for sharing because adequate storage capacity is required if the data to be shared are held in a central repository (Kevany, 1995). Yet this extra capacity can also help to avoid divergent databases if the data sets are shared at each step in the development and it reduces the likelihood of duplicated storage (Calkins and Weatherbe, 1995).

Sixth, the *introduction of standards* that enable interoperability of spatial data from different and often divergent sources, can be regarded as facilitating or as a burden owing to the changes that this may entail.

Seventh, the increased *quantity of spatial data* available to an organisation can be perceived as beneficial or a hinderance. For example, it has been suggested that if more spatial data sets can be taken into account, this would lead to better decisions. However, additional sources of spatial data need to be processed, compared and integrated with existing data sets.

Finally, *spatial data quality* has received considerable mention in the spatial data sharing literature. It is often assumed that sharing will improve the quality of spatial data (e.g. Pinto and Onsrud, 1995; MSC, 1993; Craig, 1995; Dueker and Vrana, 1995). It is argued that the wider the exposure of the data to the GIS user community, the greater the possibility of errors and gaps in the data sets being discovered. The underlying assumption is that errors will be corrected (e.g. Craig, 1995) rather than that new ones will be introduced. However, different organisations have differing quality standards and the sources of error and types of imperfection include measurement instruments, uncertainty of definition leading to irreplicable measurements, lack of documentation, distortions in physical media, GIS processing, and interpretation (Goodchild, 1995a).

Frank (1992) has suggested that organisations contemplating the sharing of spatial data need to carefully define the data quality associated with the spatial objects and attributes. Based on the US Spatial Data Transfer Standard, five dimensions of data quality can be defined for the elements of a spatial database: 1) positional accuracy, 2) attribute accuracy, 3) consistency, 4) completeness, and 5) lineage (NIST, 1992). Taking these dimensions into account, the possible outcomes of sharing for the quality of spatial data can be assessed.

Examples of the corresponding questionnaire items for these beliefs are presented in Table 5.3.

Table 5.3   Sample questionnaire items: belief about resource outcomes

a) Outcome evaluation								
Please consider the following resource-related <u>cost implications</u> of spatial data sharing.							
1. Time required to locate spatial data <u>ex-house</u>							
1	2	3	4	5	6	7	□
insignif. costs	slightly	mode-rately	fairly	quite	very	extremely significant costs	no op./ not relev.

b) Behavioural belief								
How likely is it that your organisation's engagement in spatial data sharing would result in the following resource-related cost implications?							
2. Sharing would require time to locate spatial data <u>ex-house</u>							
1	2	3	4	5	6	7	□
extrem. unlikely	slightly	mode-rately	fairly	quite	very	extrem. likely	no op./ not relev.

Note: Abbreviations used in this and any subsequent scales are due to space constraints and were not used in the original questionnaire instrument.

Attitude Domain 2: Organisational Activities

The second domain that can be considered to affect the 'attitude' toward sharing spatial data refers to organisational activities that may be positively or negatively affected by sharing spatial data across organisational boundaries. This domain has been included based on the interviews in Phase I and a survey of the literature on sharing spatial data. Three beliefs were considered important (see Table 5.4); the focus on the organisation's core activities, the usefulness of the GIS and the organisation's decision making process.

Table 5.4   Behavioural beliefs about organisational activities

Belief
1.   Focus on core activity
2.   Usefulness of GIS
3.   Decision making

First, the *core activity of an organisation* refers to the mandated operation or area of expertise of the organisation. Time and resources may be freed up by sharing spatial data to focus on the core activity since the GIS is usually only a supporting tool for the core activity of an organisation. However, it has been suggested that the

> ... efforts to make public information available to other agencies, private organisations, and individuals will overwhelm agencies and leave them less able to execute their normal mandated functions. (Epstein, 1995: 311)

This may be applicable to public and private sector organisations.

Second, sharing may have an impact on the perceived *usefulness of the GIS* in an organisation.

> Systems integration can change an existing administrative, management, or analytical environment in much the same way as the original adoption of GIS technology and can have the effect of refocusing the nature of geographic information processing and product generation within an organisation. (Dueker and Vrana 1995: 152)

Pinto and Onsrud (1995) have questioned whether the mere availability of new sources of information through sharing means that people will actually make use of this information.

Finally, it is often assumed that the *decision-making process* in a GIS-using organisation would be improved by sharing spatial data.

> The goals of GISs are to serve the information needs of both operational managers and policy and decision makers. ... A fully developed spatial data-sharing activity can only improve the ability of a GIS in meeting management goals. (Calkins and Weatherbe, 1995: 64)

Similarly, Dueker and Vrana (1995) suggest that by sharing, 'Increased resources are brought to bear on problems, and better insight is gained into difficult decision-making contexts' (pp.151-152). Yet, the availability of additional data does not necessarily lead to improved decision-making because

> ... for information to be fully utilised by an organisation, it must not simply be referred to in decision-making but must actually lead to changes in organisational values or managerial decisions (Nagel, 1988 quoted in Pinto and Onsrud, 1995: 60).

Taupier has suggested that the contribution of spatial data to decision making depends on the perceived value of the decision being made:

> The value of geographic information cannot easily be separated from its use nor can its benefits be arrived at separately from the value of the final decisions and social outcomes to which it contributes. (Taupier, 1995: 288)

While Calkins and Weatherbe (1995) have defended spatial data as a contributor to the correspondence of different decisions

> ... spatial data sharing can also ensure that decisions are made with consistent data and that decisions separated by space or time are still made with a common database. (p.66)

It can also be argued that decisions based on others' data are more likely to be questioned. Table 5.5 presents an example of the corresponding questionnaire items for these beliefs.

Table 5.5   Sample questionnaire items: belief about organisational activities

Outcome evaluation							
The way in which our organisation's spatial data sharing activities would impact the following qualitative work aspects would be ...: 1. Focus on core activities							
-3	-2	-1	0	1	2	3	☐
extrem. negative	quite	slightly	uncertain	slightly	quite	extrem. positive	no op./ not relev.

Attitude Domain 3: Strategic Position

The third domain refers to the perceived impact of sharing data upon the individual organisation's strategic position. This domain is included because many of the beliefs elicited in the initial phase of the research were related to concerns about an organisation's strategic position or its competitive position if it were to experience a loss of control over its information and ideas. Some interviewees stated that spatial data constitutes a strategic advantage which they did not want to lose through sharing. Similar concerns have been documented:

Budget constraints create situations where individuals are concerned about the viability and continuation of their programs (and jobs) and, in their own self-interest, are resistant to sharing knowledge (and data) that might make someone else's program appear more successful. (Tosta, 1995: 202)

However, others also suggested that sharing was essential to prevent users from going to competitors who may be able to provide them with the data. The issue underlying these concerns is whether an organisation perceives sharing to result in a reduction or an enhancement of its strategic position in relation to other organisations.

These strategic position-related beliefs can be operationalised in a cost/benefit-type assessment of the gains and losses that an organisation perceives will result from sharing in terms of its intellectual functioning and positioning. To this end, three groups of beliefs can be identified. These recognise the resource-based implications (first subsection), knowledge-based outcomes of sharing spatial data (second subsection) and outcomes in terms of an organisation's relationship with other organisations (third subsection). A distinction is made between the information shared in the form of spatial data sets, and tacit and explicit knowledge created and transferred in the exchange. In this way, it is possible to distinguish between the tangible effects of sharing that involve the exchange of discrete pieces information (in the form of data sets) and more intangible effects of sharing ideas and knowledge and the extent to which these are perceived as learning opportunities or threats.

Loss of control over spatial data The first group of beliefs under strategic position that may affect the 'attitude' toward sharing are beliefs about the tangible aspects of sharing regarding the loss of control over spatial data. Kevany describes different methods of controlling spatial data across organisational boundaries:

The information may be controlled through a central management mechanism. The information may be controlled by individual "owning" nodes. The control may be distributed through all organisations in a distributed configuration. Control may also be established on an ad hoc basis to suit particular needs of each situation. (Kevany, 1995: 88)

Many interviewees said that they were concerned about a loss of control over their organisation's spatial data. In order to operationalise this belief, the resource dependence approach is employed. Resource dependence theory attempts to assess the extent of dependence on others by evaluating the amount of control other parties have over a required resource. These evaluations of control serve here to measure the extent to which respondents perceive sharing as a *loss of control* over their information and how this influences their 'attitude' towards sharing. Four aspects of control over information were considered; copyright, access to spatial data, control of the use of spatial data, and the enforcement of rules regarding spatial data (see Table 5.6).

Table 5.6   Behavioural beliefs about loss of control over spatial data

Belief
1. Copyright
2. Access
3. Use
4. Enforcement of rules

Pfeffer and Salancik (1978) have identified four elements that together make up the extent to which an organisation has discretion over the allocation and use of a resource possessed by another social actor. These are 1) possession, in terms of ownership or ownership rights, 2) regulated access without necessarily owning the resource, 3) actual use and control of the resource, and 4) the ability to rule or regulate possession, allocation and use of the resource and enforcement of legislation. Pfeffer and Salancik developed these concepts to analyse the dependence of an organisation on one (or more) other organisation(s) for a particular resource. What is assumed to be measured is the level of control an organisation has over a resource that it *does not* possess. It is also possible to employ these concepts for evaluating the possible outcomes of losing control over a resource that an organisation *does* possess.

The resource in question in this case is the spatial data being shared and the perspective shifts to the discretion of an organisation over the allocation and use of the spatial data it possesses. The four elements of control can be operationalised as follows. Indirect means of discretion over a resource are presented by ownership and ownership rights (Pfeffer and Salancik, 1978). Evaluating the possible outcomes of losing this element of control, organisations are considering whether not having copyright for the spatial data they make available to other organisations presents an opportunity or a threat. The second element of control - regulated access without necessary ownership - can be operationalised by evaluating the implication of unregulated access to an organisation's spatial data by other organisations. The third element of control - actual use and control of the resource - can be implemented by asking individuals within an organisation to consider the use of its spatial data by other organisations. Evaluating the loss of the final component - the ability to make rules or to regulate possession, allocation and use of the resource and enforcement of legislation - organisations are considering the effects of not being able to enforce any rules regarding the ownership of, access to, and use of, their spatial data. The evaluation of these possible outcomes related to losing control over information is combined with an assessment of the likelihood of these outcomes occurring.

Together, these aspects of an organisation's strategic position may be expected to affect the 'attitude' towards spatial data sharing.

Knowledge creation The second group of beliefs considered as part of the strategic position outcomes of sharing spatial data consists of the perceived intangible effects of sharing spatial data. The interviews revealed beliefs about intangible consequences of spatial data sharing such as the chance to gain new insights or the fear of ideas being appropriated.

The relation of these intangible aspects to the concrete exchange of spatial data needs to be clarified. Spatial data are discrete facts. Data are transformed into information when they are contextualised, categorised, analysed, or summarised and, in turn, information can be transformed into knowledge, for example through comparison or connection with other information (Davenport and Prusak, 1998; Haywood, 1995). Knowledge is defined as follows;

> Knowledge is a fluid mix of framed experience, values, contextualised information and expert insight that provides a framework for evaluating and incorporating new experiences and information. (Davenport and Prusak, 1998: 5)

This highlights the contribution of new experience and insights to knowledge. The appropriate context within which to consider the beliefs about gaining new insight or diluting ideas is the creation of knowledge. Knowledge has been recognised as a crucial element of an organisation's competitive advantage (Nonaka *et al*, 1996).

Based on Polanyi's (1966) distinction between tacit and explicit knowledge, Nonaka (1994 and Nonaka *et al.* 1996) propose four different modes of knowledge creation. *Socialisation* denotes the sharing of experiences and creation of tacit knowledge through shared mental models; *externalisation* involves articulating tacit knowledge into explicit concepts; *combination* is the systematisation of concepts into a knowledge system; and *internalisation* is the absorption of experiences from socialisation, externalisation and combination into tacit knowledge.

The basic assumption is that knowledge is created through the social interaction between tacit and explicit knowledge. This is referred to as a knowledge spiral:

> Thus, organizational knowledge creation is a spiral process, starting at the individual level and moving up through expanding communities of interaction, which crosses sectional, departmental, divisional, and organizational boundaries. (Nonaka et al., 1996: 842)

The scope of the knowledge creation mechanisms considered by Nonaka *et al.* (1996) ranges from the individual to inter-organisational interactions and takes into account the interaction of an organisation with the environment (Nonaka, 1994). It applies within and across organisations. It can be utilised for the inter-organisational setting considered in this study to address the beliefs about intangible outcomes of spatial data sharing for an organisation's strategic position.

The specific beliefs considered under knowledge creation are listed in Table 5.7.

Table 5.7   Behavioural beliefs about knowledge creation

Belief
1.   Explicit to explicit knowledge
2.   Explicit to tacit knowledge
3.   Tacit to tacit knowledge
4.   Tacit to explicit knowledge

The different modes of knowledge creation are particularly suitable for capturing the knowledge transfer and creation accompanying the exchange of spatial data across organisational boundaries. The different knowledge creation modes of spatial data sharing may be perceived as a threat or as an opportunity for an organisation's strategic position (see Figure 5.2).

First, in the context of sharing spatial data across organisational boundaries, externalisation refers to triggering dialogue and collective reflection across organisational boundaries, such as joint problem solving, and involves articulating, making explicit and comparing spatial data-related concepts. Second, the internalisation of explicit knowledge to create tacit knowledge can take place by allowing access to ideas that are codified in spatial data, for example, as thematic layers, and their translation, combination and application in different organisations. Third, socialisation, the creation of tacit knowledge from tacit knowledge, can take place when the means to observe spatial data skills in personal interaction and to create shared mental models and experiences across organisational boundaries are provided. Finally, the combination of new and existing spatial data sets from different organisations enables the creation of new explicit knowledge.

	to	
	Tacit knowledge	*Explicit knowledge*
Tacit Knowledge from	***Socialisation*** Provide the means to observe spatial data skills in personal interaction and/or to create shared mental models and experiences across organisational boundaries.	***Externalisation*** Trigger dialogue and collective reflection across organisational boundaries (e.g. joint problem solving) and articulate, make explicit and compare spatial data-related concepts.
Explicit Knowledge	***Internalisation*** Allow access to ideas that are codified in spatial data (e.g. as thematic layers) and their translation, combination and / or application in different organisations.	***Combination*** Enable the combination of new and existing spatial data sets from different organisations to create new knowledge.

Figure 5.2   Opportunities for knowledge creation in spatial data exchanges
Source: based on Nonaka (1994)

By considering each knowledge creation element separately, it is possible to distinguish which knowledge creation mode of spatial data sharing poses the greatest threat or opportunity to organisations. Organisations that value these knowledge creation chances resulting from sharing and related personal interactions regard them as opportunities for their organisation whereas those organisations that are worried about losing ideas feel threatened by them; accordingly, these beliefs may be expected to influence the 'attitude' towards spatial data sharing.

Inter-organisational relations The final dimension of strategic position beliefs that may be considered to affect 'attitude' is constituted by outcomes in terms of an organisation's relationship with other organisations. This relationship dynamic also deals with intangible aspects of spatial data sharing. Two beliefs are included concerning the interdependence among organisations and the redistribution of influence among organisations (see Table 5.8).

Table 5.8   Behavioural beliefs about inter-organisational relations

Belief
1. Interdependence
2. Redistribution of influence

Inter-organisational relationships constitute one of the resources that contribute to an organisation's strategic position (Porter, 1991). Inter-organisational dynamics are also a central element of the framework of Azad and Wiggins (1995) for managing spatial data sharing (see Chapter 3). However, they have an inherently negative view of inter-organisational relations;

> ... any geographic data-sharing effort, no matter how low in intensity, will involve a redefinition of existing tasks / structures/ and / or the redistribution of relative power and influence in organizational environment. (Azad and Wiggins, 1995: 23)

Their view focuses on the loss of an organisation's autonomy:

> As such, any organization's decision to engage in inter-organizational relations (data sharing) is primarily determined by the autonomy "calculations" of the outcomes; that is, each organization will attempt to minimize its loss of autonomy and therefore interdependency with other organizations. (Azad and Wiggins, 1995: 41)

Alter and Hague's (1993) review of the inter-organisational co-operation literature has identified many different benefits and costs of inter-organisational, co-operative relationships, of which loss of autonomy is only one. The relationship of an organisation with other organisations and the organisation's position within a network can contribute to its strategic advantage (Galaskiewicz and Zaheer 1999). Furthermore, over time, the costs and benefits of inter-organisational co-operation are expected to shift (Harrigan and Newman, 1990).

In order to capture the perception of the outcomes of sharing for an organisation's strategic position in terms of inter-organisational relations, two aspects of the dynamic are assessed: the interdependence of organisations and the redistribution of influence. The constructs included in the TPB provide a basis to measure the perception of organisations as to whether interdependence and redistribution of influence among organisations present opportunities or threats, and how likely these are to result from sharing spatial data across organisational boundaries.

Together, beliefs about inter-organisational relations, knowledge creation, and control of spatial data make up the perception of the implications of spatial data sharing for an organisation's strategic position. Examples of the corresponding questionnaire items for these beliefs are given in Table 5.9.

Table 5.9   Sample questionnaire items: belief about strategic position outcomes

a) Outcome evaluation							
Spatial data sharing can create threats and opportunities for the strategic position of all parties involved - the strategic position of your organisation and that of the organisations you may be sharing with. Please indicate how you value the following consequences of sharing spatial data for your organisation's strategic position. 1. If sharing were to enable the combination of new and existing spatial data sets from different organisations to create new knowledge, in terms of our organisation's strategic position this would present a ...:							
-3	-2	-1	0	1	2	3	☐
extreme threat	quite	slight	uncertain	slight	quite	extreme opportu-nity	no op./ not relev.
b) Behavioural belief							
How likely is it that spatial data sharing would have the following consequences for the strategic position of your organisation? 1. Sharing would enable the combination of new and existing spatial data sets from different organisations to create new knowledge.							
-3	-2	-1	0	1	2	3	☐
extrem. unlikely	quite	slightly	uncertain	slightly	quite	extrem. likely	no op./ not relev.

Attitude Domain 4: Social Outcomes

The last domain to be taken into account which may affect the 'attitude' towards spatial data sharing, is that dealing with social outcomes. As discussed in Chapter 3, morality can be considered in terms of the moral implications, or social outcomes, of performing a particular behaviour. Since the interviews in Phase I elicited moral beliefs, two specific beliefs were included; integrated development planning and the distribution of benefits to society at large (see Table 5.10).

Table 5.10   Behavioural beliefs about social outcomes

Belief
1.   Integrated development planning
2.   Distribution of benefits to society at large

Integrated development planning was mentioned frequently by interviewees in the interviews in Phase I. It denotes a specific initiative by the South African government targeting the local and provincial government levels. These levels of government have to justify their activities in term of 'integrated development' and provide specific plans forecasting how the lives of citizens would be improved (Savage *et al.*, 1999; Liebenberg, 1999). Benefits to society at large are proclaimed by spatial data infrastructure initiatives (Nebert, 2000; NSIF, 1998) and in the literature, for example, 'From the perspective of society at large, the more users [of geographic information] there are the greater the value of the benefits that will be generated' (Taupier, 1995: 289).

An example of the corresponding questionnaire items for these beliefs is given in Table 5.11 below. As these beliefs are inherently positive, outcome evaluations are not included.

Table 5.11   Sample questionnaire items: belief about social outcomes

b) Behavioural belief							
Our organisation's spatial data sharing activities would result in the following social benefits:							
1. Integrated development planning							
-3	-2	-1	0	1	2	3	☐
extrem. unlikely	quite	slightly	uncertain	slightly	quite	extrem. likely	no op./ not relev.

Summary of Attitude Determinants

The domains underlying the 'attitude' towards sharing spatial data - resources, organisational activities, strategic position and social outcomes - have been introduced. Together, they may be expected to influence an organisation's 'attitude' towards spatial data sharing which, in turn, may influence the organisation's willingness to share. The 'attitude' domains are depicted in Figure 5.3.

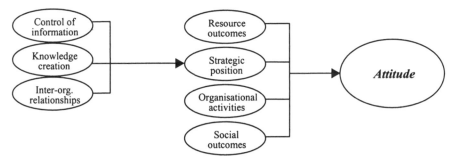

Figure 5.3   Determinants of attitude towards spatial data sharing

An example of the questionnaire items at the level of the domains belonging to the 'attitude' component is provided in Table 5.12.

Table 5.12   Sample questionnaire items: attitude domain-level (resource outcomes)

In terms of our organisation's resources (time, money, labour, data storage, quality and quantity of spatial data), spatial data sharing would result in ...:
1. at the moment

-3	-2	-1	0	1	2	3	☐
extrem. large costs	quite	slightly	uncertain	slightly	quite	extrem. large benefits	no op./ not relev.

2. in the next month - 2 years

-3	-2	-1	0	1	2	3	☐
extrem. large costs	quite	slightly	uncertain	slightly	quite	extrem. large benefits	no op./ not relev.

Table 5.13 presents an example of the 'direct' measures of 'attitude' towards spatial data sharing.

Table 5.13   Sample questionnaire items: direct measure of attitude

The consequences of sharing spatial data for our organisation's resources, qualitative work aspects, and strategic position, and in terms of social benefits would be ...:
1. at the moment

-3	-2	-1	0	1	2	3	☐
extrem. negative	quite	slightly	uncertain	slightly	quite	extrem. positive	no op./ not relev.

2. in the next month - 2 years

-3	-2	-1	0	1	2	3	☐
extrem. negative	quite	slightly	uncertain	slightly	quite	extrem. positive	no op./not relev.

Social Pressure to Engage in Spatial Data Sharing

'Social pressure' refers to the perceived pressure by important referents on key individuals with an organisation to engage, or not, in the behaviour in question. This section relies heavily on the beliefs elicited during the interviews in Phase I because there was little in the spatial data sharing literature that was helpful for identifying the important referents for spatial data sharing. This illustrates that the use of the TPB as an organising framework for the analysis of spatial data sharing creates opportunities to include aspects that have been neglected by others. Five domains that serve as proxies for 'social pressure' were established: pressures from the GIS community, from within the organisation, from institutions and the market, and moral norms. Each 'social pressure' domain is made up of normative beliefs about the important referents (individuals or groups). Normative beliefs are assessed by determining the likely approval or disapproval of spatial data sharing by the referents and the general motivation to comply with the perceived views of each referent.

Social Pressure Domain 1: GIS Community Pressure

The first domain to be considered under 'social pressure' is the GIS community. This domain refers to beliefs about the expectations of different groups within the GIS community and whether these groups are likely to approve or disapprove of the organisation's engagement in spatial data sharing. The distinct groups within the GIS community have been identified by Abbott (1996) and were confirmed during the interviews in Phase I (see Table 5.14).

Table 5.14   Normative beliefs about GIS community pressure

Belief
1.   GIS users in local authorities
2.   GIS users in provincial government
3.   GIS users in national government
4.   GIS departments in para-statal organisations
5.   Private sector GIS users
6.   NGO GIS users
7.   Academic research institutions

Examples of the corresponding questionnaire items for these beliefs are given in Table 5.15.[27]

---

[27] The same format applies to the questionnaire items for the beliefs in the domains of market pressure, institutional pressure and organisational pressure.

Table 5.15   Sample questionnaire items: belief about GIS community pressure

a) Normative belief
The following people think that our organisation's engagement in spatial data sharing would be ...:
1. GIS departments in local authorities

-3	-2	-1	0	1	2	3	□
extrem. negative	quite	slightly	uncer-tain	slightly	quite	extrem. positive	no op./ not relev.

b) Motivation to comply
In general, our organisation very much wants to do what the following people want us to do ...:
1. GIS departments in local authorities

1	2	3	4	5	6	7	□
not at all	slightly	moder-ately	fairly	quite	very	extremely (very much)	no op./ not relev.

Social Pressure Domain 2: Market Pressure

The second domain with respect to 'social pressure' is market pressure. A market for spatial data is developing and various stakeholders are involved. These include the commercial spatial data brokers, public and private providers, and GIS suppliers who may be branching out into data supply (see Table 5.16). Market pressure refers to beliefs about their likely approval or disapproval of spatial data sharing.

Table 5.16   Normative beliefs about market pressure ·

Belief
1. Commercial spatial data brokers
2. Public spatial data providers
3. Private spatial data providers
4. GIS suppliers

Social Pressure Domain 3: Institutional Pressure

The third domain that may affect the 'social pressure' to engage in spatial data sharing is institutional pressure. This domain includes beliefs about the likely approval or disapproval of institutional referents and pressures with regard to the organisation's engagement in spatial data sharing. The interviews identified three specific referents: the NSIF in South Africa, spatial data agreements, and politicians (Table 5.17).

Table 5.17   Normative beliefs about institutional pressure

Belief
1.   NSIF
2.   Spatial data agreements
3.   Politicians

A central feature in the South African set up is the NSIF in the Department of Land Affairs established to facilitate spatial data sharing. It focuses mainly on the public sector but also involves the private sector. Secondly, spatial data agreements that are already in place or are in the process of being set up can present a source of pressure for an organisation either to share or not to share (e.g. exclusive agreement). Finally, politicians may act as advocates of spatial data sharing.

Social Pressure Domain 4: Organisational Pressure

The fourth domain contributing to the perception of 'social pressure' refers to organisational pressures. Four perceived sources of organisational pressure were mentioned in the interviews in Phase I with regard to the influence they may have on an organisation's sharing behaviour. These are other departments in the organisation, the management of the organisation, particular individuals within the organisation, and the organisation's mandate (see Table 5.18).

Table 5.18   Normative beliefs about organisational pressure

Belief
1.   Other departments within the same organisation
2.   Management of the organisation
3.   Individual champions for sharing
4.   Mandate (organisational goals/mission)

The different levels within the organisational structure - from individuals to departments to management - can be conceived as referents for spatial data sharing. According to the interviews in Phase I and the spatial data sharing literature, individuals can be an important referent when they act as champions for spatial data sharing, demonstrating leadership and commitment (Kevany, 1995; Craig, 1995; Bamberger, 1995; Obermeyer, 1995; Masser and Campbell, 1995). Finally, an organisation's mandate can present another source of organisational pressure:

> The focus and breadth of the goal or mission of the organizations involved will impact sharing. Some organizations have a specific function, such as issuing permits or providing water service. Others have as their goal or mission service to other organizations, such as providing information services or mapping services. (Kevany, 1995: 87)

Social Pressure Domain 5: Moral Norms

The final domain affecting 'social pressure' is moral norms. As discussed in Chapter 3, the moral norms domain provides one way to address morality beliefs. In this domain, the morality beliefs about integrated development planning and society at large are conceptualised as moral norms. As such, they address the extent to which organisations perceive moral pressure or an obligation to share data for the sake of integrated development planning and society at large (see Table 5.19).

Table 5.19   Perceived moral norms

Belief
1.   Integrated development planning
2.   Society at large

Examples of the corresponding questionnaire items for these beliefs are given in Table 5.20 below.

Table 5.20   Sample questionnaire items: belief about moral norms

a) Normative belief							
The following social beneficiaries suggest that our organisation's engagement in spatial data sharing would be desirable. 1. Integrated development planning							
-3	-2	-1	0	1	2	3	☐
strongly disagree	quite	slightly	uncertain	slightly	quite	strongly agree	no op./ not relev.
b) Motivation to comply							
How willing is your organisation to engage in spatial data sharing just for the sake of the following beneficiaries? 1. Integrated development planning							
1	2	3	4	5	6	7	☐
not at all	slightly	moder-ately	fairly	quite	very	extremely (very much)	no op./ not relev.

Summary of Social Pressure Determinants

This section has presented the domains that may be expected to underlie the 'social pressure' to share spatial data - GIS community, organisational, institutional, and market pressures, and moral norms. The beliefs grouped in these domains are expected to affect an organisation's perception of 'social pressures' to share spatial data. The 'social pressure' domains are depicted in Figure 5.4.

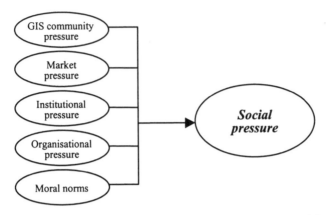

Figure 5.4   Determinants of social pressure to engage in spatial data sharing

An example of the questionnaire items at the level of the domains of the 'social pressure' component is provided in Table 5.21.

Table 5.21   Sample questionnaire items: social pressure domain-level (GIS community pressure)

Overall, the pressure of the GIS community regarding our organisation's sharing activities is likely to be ...:

1. at the moment

1	2	3	4	5	6	7	☐
extrem. weak	slightly	moder- ately	fairly	quite	very	extrem. strong	no op./ not relev.

2. in the next month - 2 years

1	2	3	4	5	6	7	☐
extrem. weak	slightly	moder- ately	fairly	quite	very	extrem. strong	no op./ not relev.

Table 5.22 presents an example of the 'direct' measures of 'social pressure' to share spatial data.

Table 5.22   Sample questionnaire item: direct measure of social pressure

People and institutions that are important to the success of our organisation think we should engage in spatial data sharing.

1. at the moment

-3	-2	-1	0	1	2	3	☐
strongly disagree	quite	slightly	uncertain	slightly	quite	strongly agree	no op./ not relev.

2. in the next month - 2 years

-3	-2	-1	0	1	2	3	☐
strongly disagree	quite	slightly	uncertain	slightly	quite	strongly agree	no op./ not relev.

Perceived Control over Spatial Data Sharing

The TPB suggests that when research is concerned with behaviour that may not be completely under an individual's volitional control, it is necessary to take into account control beliefs that may affect what is called 'perceived control'. The 'perceived control' component is intended to measure how easy or difficult the performance of spatial data sharing is for individuals within organisations due to the presence or absence of the requisite resources and opportunities. It serves as a means to elicit the perceived degree of control of an organisation over the various aspects of spatial data sharing. The control beliefs influencing 'perceived control' capture the perceived presence or absence of facilitating or inhibiting factors for spatial data sharing such as the requisite resources and opportunities. Each control belief needs to be assessed with respect to both the perceived control over, and the perceived importance of, the factor.

There is a wide range of possible sharing arrangements (see Calkins and Weatherbe, 1995 for a taxonomy of spatial data sharing). No single co-operation mechanism or sharing arrangement is prescribed by the model of spatial data sharing developed in this study and not necessarily all of the control aspects may be considered by any one organisation. Hence, the importance organisations give to these control aspects may differ substantially. However, this is a crucial indicator to understand the conceptualisation of spatial data sharing in different organisational settings.

Two groups of control factors can be distinguished; internal and external control factors (Ajzen and Madden, 1986; Ajzen 1988). Usually, the *internal* control factors, comprising skills, abilities, knowledge and adequate planning, are considered with respect to the individual respondent. However, since the focus of this study is on organisational willingness to share spatial data, the skills and abilities that are considered here are under the organisation's control, rather than only those (or distinguishing between the respondent and others within the organisation) of the individual.

External control factors include the extent to which performing the behaviour depends on appropriate opportunities and on the co-operation of others (Ajzen and Madden, 1986). Applied to spatial data sharing, dependence on others is considered with respect to dependence on sharing partners.

Internal Factors

The internal factors are organised into the following four domains; technical skills, interpersonal skills, control over resources for sharing, and dependence on, and control over, spatial data.

Perceived control domain 1: Technical skills   The first domain underlying 'perceived control' is the technical skills domain. From the interviews in Phase I and from the literature on spatial data sharing, it is apparent that a range of technical skills and activities is involved in spatial data sharing. Three specific areas can be identified which, together, make up the technical skills domain; spatial data related activities, metadata skills, and general information technology skills.

1. Spatial data skills
The first area of technical skills to be considered is spatial data-related skills. It refers to beliefs about control over spatial data sharing in terms of spatial data skills. These include assessing the quality of spatial data, coping with different format and standards, and selecting and integrating spatial data. The specific beliefs are listed in Table 5.23. The evaluation of each belief concerns the organisation's perception of whether each skill is important for sharing spatial data with other organisations and whether each skill is easy or difficult for the organisation. This evaluation can be assumed to generate a high or low level of 'perceived control' over spatial data sharing. Each belief is briefly discussed below.

Table 5.23   Control beliefs about spatial data skills

Belief
1. Assessing quality of spatial data
2. Handling different formats of spatial data
3. Mastering different standards
4. Selecting spatial data for exchange
5. Integrating different spatial data sets

First, as discussed above, the *quality of spatial data* can be defined in terms of five dimensions following the US Spatial Data Transfer Standard (NIST, 1992). The ability to assess the quality of spatial data entails the appraisal of positional accuracy, attribute accuracy, consistency, completeness, and lineage for the elements of a spatial database or subset.

Second, the possibility of having to deal with *different formats of spatial data* arises from the diversity of the GIS market that is made up of a range of GIS software suppliers. Different GIS packages rely on differing, although not exclusive, spatial data formats and the exchange of spatial data across different GIS platforms demands the ability to convert spatial data databases or subsets from one format to another.

Third, *standards* form a central component of the spatial data sharing initiatives (see Chapter 2). While many organisations share the need for particular spatial data sets, exact specifications and priorities differ (Bamberger, 1995) and often data have been captured to different standards. As Calkins and Weatherbe (1995) have pointed out: 'The major difference between spatial data sharing and data sharing in other contexts is the need to have common definitions and standards for the spatial data' (Calkins and Weatherbe, 1995: 69).

The definition of these standards is complex and stems from the different levels of detail required by different applications and the different types and structure of data currently used (Alfelor, 1995). While the agreement and development of standards is not always straightforward, for many users the implementation of these standards implies that application programs will have to be changed or adjusted with respect to their individual data-handling functions and interfaces (Alfelor, 1995). Mastering different standards for spatial data sharing entails both spatial data standards and attribute data standards such as feature type classifications (NSIF, 1998).

Fourth, spatial data exchanges can range from a single observation, summary or aggregate observations to selected subsets and themes to the exchange of an entire spatial database (Calkins and Weatherbe, 1995). In order to share these observations, subsets or thematic layers, they have to be *selected* and extracted from the spatial data base.

Finally, using somebody else's data can add complexity to spatial data processing since it is often easier to work with one's own captured data. *Integrating different spatial data sets* requires the ability to convert and manipulate the data, verify its geometric and attribute quality, maintain its consistency with related data, and update it over time (Evans and Ferreira, 1995).

2. Metadata skills

The second area of technical skills considered to affect the 'perceived control' over sharing spatial data refers to metadata-related skills that may be perceived as being easy or difficult and important or not for sharing spatial data across organisational boundaries.

Metadata have been labelled as 'well-defined measures of uncertainty' (Goodchild, 1995a: 421). This information about data describes the content, ancestry and source, quality, data base, schema, and accuracy of spatial data. (MSC, 1993). Its purpose is to 'allow the potential user of spatial data to understand that data's fitness for use' (Onsrud and Rushton, 1992: 12). Metadata constitute one of the basic components of the spatial data infrastructure initiatives and there are a number of related skills that are considered under 'perceived control' (see Table 5.24).

Table 5.24   Control beliefs about metadata

Belief
1. Interpret metadata
2. Use metadata interfaces / databases
3. Capture metadata
4. Apply metadata standards
5. Maintain / update metadata

First, interpreting metadata entails using the elements of metadata to determine whether a dataset (or database) is adequate for the intended purpose. Second, metadata is typically accessible through specific interfaces or in a database environment. Third, metadata about spatial data needs to be captured. Fourth, particular metadata standards may also need to be applied. Finally, metadata requires maintenance and updating in parallel with changes made to the spatial data it describes.

3. Information technology skills
The final area of technical skills   assumed to determine 'perceived control' are information technology skills. Several general information technology skills form part of the skills and activities of spatial data sharing (see Table 5.25).

Table 5.25   Control beliefs about IT skills

Belief
Database administration
Using the Internet to locate spatial data sources
Using the Internet to distribute spatial data
Transfer spatial data to/from different media

First, database administration skills are necessary to control spatial and attribute databases. Next, using the Internet to *locate* spatial data sources and to *distribute* spatial data entails accessing and searching multiple data directories in diverse locations over computer networks (MSC, 1993). Finally, spatial data sharing can also entail the transfer of spatial data to and from different media such as CD-ROM. Examples of the corresponding questionnaire items for these beliefs are given in Table 5.26.[28]

---

[28] The same format applies to the questionnaire items for the beliefs about negotiation skills as well as networking and teamworking skills in the interpersonal skills domain. This domain is presented in the next subsection.

Table 5.26   Sample questionnaire items: belief about technical skills

a) Control belief							
For your organisation, how difficult are the following spatial data-related activities? 1. Handling different formats of spatial data?							
-3	-2	-1	0	1	2	3	
extrem. difficult	quite	slightly	uncertain	slightly	quite	extrem. easy	no op./ not relev.
b) Perceived importance							
How important are the following spatial data-related skills for sharing spatial data? 1. Handling different formats of spatial data							
1	2	3	4	5	6	7	
not at all important	slightly	moder- ately	fairly	quite	very	extrem. important	no op./ not relev.

Perceived control domain 2: Interpersonal skills The second domain underlying 'perceived control' is the interpersonal skills domain. The interpersonal skills have been grouped here into the areas of networking and teamworking, negotiation skills, and past experience. The interpersonal skills domain has been included based on the evidence from the interviews in Phase I and the literature on sharing spatial data.

1. Networking and teamworking skills
The first area of interpersonal skills considered to influence 'perceived control' over sharing spatial data is networking and teamworking skills. Alter and Hague (1993) have defined networking as the act of creating and/or maintaining a cluster of organisations for the purpose of exchanging. Networking skills refer to the ability to establish and maintain personal relationship across organisations and can be considered to consist of several activities (see Table 5.27).

Table 5.27   Control beliefs about networking and teamworking skills

Belief
  Establish / foster contacts
  Identify meeting opportunities
  'Keep finger on the pulse of a network'
  Collaborate with other
  Multi-disciplinary teamwork

First, contacts need to be established and fostered in order to build a network of contacts. Second, opportunities to meet with other parties who are also working with spatial data need to be identified. Third, the belief that keeping a finger on the pulse of a network in terms of what others are working on and letting them know that your own work is important for sharing resonates with Granovetter (1972). Granovetter has pointed out that the maintenance of weak ties of networks may be one of the most important aspects of meeting opportunities. His concern was mainly with the strength of the ties of network members.[29] Following Granovetter's argument, the importance of weak ties lies in the fact that crucial information about who is doing what in terms of spatial data generation and application is passed on. Nevertheless, according to the interviewees in Phase I, everybody 'wants stuff' but they do not want to give it out or even let others know what they have.

Finally, teamworking skills are assessed in terms of the ability to collaborate with other and to work across different disciplines.

2. Negotiating skills

The second area of interpersonal skills considered likely to affect 'perceived control' over spatial data sharing is negotiating skills. Sharing arrangements can range from a formal agreement to address sharing mechanisms in detail, a general formal contract, informal working relationships, no agreement at all, or ad hoc arrangements on a case-by-case basis (Calkins and Weatherbe, 1995; Kevany, 1995). The characteristics of a sharing arrangement are agreed upon by the participating organisations through a process of negotiation. In order to measure the extent to which organisations have negotiation skills, the ease or difficulty of negotiating the different characteristics of a sharing arrangement are considered (see Table 5.28).

Table 5.28   Control beliefs about negotiation skills

Belief
1.   Win-win arrangement
2.   Pricing of spatial data
3.   Data ownership agreement
4.   Liability agreements

First, one aspect of the arrangement is whether there is leadership of one organisation, equality among participants, or groupings of lead and sharing organisations (Kevany, 1995). A *win-win relationship* exists where a sharing arrangement benefits all parties.

Second, the *cost of spatial data* is another aspect of the sharing arrangement. It has been argued that very high prices can limit sharing where the costs are high enough to discourage most users (Craig, 1995). By preventing people from utilising spatial data that already exist, they have to capture their own data or purchase them elsewhere and may end up with a poorer product. Alternatively, 'pricing for access' to spatial data might be considered, but even the marginal cost of providing access can be considerable (King, 1995). Whether sharing involves a licence fee, prices reflecting the marginal cost of reproduction, barter or no charge at all, the agreement of the price is a central aspect of the negotiation of the sharing arrangement.

---

[29] Consideration is given to the content of the exchange in the section *Perceived control domain 4: spatial data position* later in this chapter.

Third, *data ownership agreements,* such as copyright, constitute the third aspect of a sharing arrangement that needs to be negotiated. Different sources of data may have different copyright provision and these may be difficult to resolve due to differences between the parties involved (Kevany, 1995).

Finally, *liability agreements* form the last negotiation point of sharing arrangements that is considered here. Onsrud (1995) suggests that '... basic liability principles mandate that some level of acceptable performance in the delivery of services or the quality of products is required' (Onsrud 1995: 300).

It has been argued that liability is the responsibility of the data providers (NSIF, 1998b). Nevertheless, it needs to be negotiated as part of a sharing agreement.

3. Past experience

The final area of interpersonal skills that is likely to influence 'perceived control' is past experience. The beliefs assessed under 'perceived control' may be assumed to reflect past experience as well as anticipated impediments and obstacles that people take into account when responding to questions (Ajzen, 1991). Therefore, the degree to which organisations perceive having control over spatial data sharing may be informed by their past experience with sharing. No specific questions on the *importance* of past experience were developed for inclusion in the survey instrument. Nevertheless, questions were included specifically asking respondents about the *extent* of their past experience (see Table 5.29).

Table 5.29   Control beliefs about past experience

Belief
1.   Past involvement in spatial data sharing with other organisations
2.   Bad experiences with spatial data sharing with other organisations
3.   Clear overview of the consequences of spatial data sharing with other organisations

In his application of the TPB to assess the willingness of organisations to change, Metselaar (1997) operationalised past experience with respect to past involvement in the behaviour, bad experiences in carrying out the behaviour, and a clear overview of the consequences of the behaviour, that is assumed to be based on experience of carrying out the behaviour.

Applied to sharing spatial data with other organisations, past experience can be measured by using as a proxy the extent to which respondents have had involvement in spatial data sharing with other organisations, the extent of bad experiences with spatial data sharing with other organisations, and a clear overview of the consequences for their organisation of spatial data sharing with other organisations. An example of the corresponding questionnaire items for these beliefs is presented in Table 5.30.

Table 5.30   Sample questionnaire items: belief about past experience

a) Control belief							
"In the past, our organisation has been heavily involved in spatial data sharing with organisations."							
-3	-2	-1	0	1	2	3	
strongly disagree	quite	slightly	uncertain	slightly	quite	strongly agree	no op./ not relev.

Perceived control domain 3: resource control The third domain that is likely to affect 'perceived control' is resource control. The beliefs in this domain refer to the absence or presence of requisite resources for sharing spatial data across organisational boundaries. The set of resources considered here is similar to that in the resource domain affecting 'attitude' of the model of spatial data sharing (see section *Attitude domain 1: Resources* in this chapter) where respondents are asked to evaluate the outcomes of sharing for their resources. In contrast, in this domain we are interested in the extent to which respondents perceive that they have control over the resources they consider necessary for spatial data sharing. GIS packages and computer equipment are taken as given, as the use of GIS is a prerequisite for inclusion of the organisations in the study. Several resources may be required for spatial data sharing (see Table 5.31).

Table 5.31   Control beliefs about resources

Belief
1. Sufficient Staff
2. Time
3. Availability of funding
4. Organisational guidelines

Trained staff must be available (Sperling, 1995), both with respect to their skills and the time that can be allocated to spatial data sharing activities. Sharing activities can also imply costs and, therefore, funding resources for these activities need to be considered. Kevany suggests that funding can take many forms such as finance by a single lead organisation, shared costs in a balanced manner across multiple organisations, or shared costs at different levels across organisations (Kevany, 1995).

The belief that organisational guidelines, such as a clear spatial data policy or procedures, are an important resource was raised during the interviews in Phase I. Otherwise, there can be uncertainty over whether sharing is allowed. The NSIF is a proponent of setting a clear policy for information sharing in which the responsibilities are spelled out (NSIF, 1998b).

Examples of the corresponding questionnaire items for these beliefs are given in Table 5.32 below.

Table 5.32   Sample questionnaire items: belief about resource control

a) Control belief							
In our organisation, the extent to which we have control over the following resources is ...:							
1. Availability of funding							
1	2	3	4	5	6	7	☐
(not at all) extrem. small	slightly	moder- ately	fairly	quite	very	extrem. large	no op./ not relev.
b) Perceived importance							
How important are the following resources for sharing spatial data?							
1. Availability of funding							
1	2	3	4	5	6	7	☐
not at all important	slightly	moder- ately	fairly	quite	very	extrem. important	no op./ not relev.

Perceived control domain 4: spatial data position The fourth domain that may be expected to influence 'perceived control' over sharing spatial data is the spatial data position of an organisation in terms of the dependence on external, and control over internal, spatial data. Although spatial data can be regarded as an organisational resource, it is the object of the behaviour under study and therefore it is not considered among the necessary resources for sharing (see section *Perceived control domain 3: Resource control* in this chapter) but is considered, instead, in this separate domain. Furthermore, the dependence on others to perform the behaviour is considered in this chapter in section *Perceived control domain 5: Dependence on others* in terms of finding sharing partners. Finding sharing partners differs from the dependence of an organisation on others for spatial data which is expected to increase the willingness to share. The focus in this section is on dependence on the target of the behaviour (i.e. spatial data) rather than on dependence on others to perform the behaviour (i.e. having to find sharing partners). As the operationalisation of these beliefs is not straightforward, this section draws on resource dependence theory.[30]

According to Kevany, there are two perspectives to consider:

Level of Dependence from Sharer Perspective
The potential organizational dependence, when viewed from the organizations that are not the leader but are sharing data, may range from high to no dependence, though the response may be to move away from the dependence of sharing toward information independence.
...

---

[30] However, the length of this section does not imply increased importance compared to other domains.

Level of Dependence from Lead Organization Perspective
The dependence of organizations on the shared information or GIS is important to the level of information sharing. Where there is a lead organization with primary control, its perception of the level of dependence of the other organizations may encourage it to facilitate sharing. Its perception of the level of dependence may range from a high level of dependence to no dependence at all. (Kevany, 1995: 89-90)

What Kevany has termed 'sharer' can be designated as the recipient organisation of spatial data, while Kevany's 'lead organisation' can be designated as the organisation making spatial data available to others. Kevany suggests that two perspectives need to be taken into account: giving and receiving spatial data. These are both aspects of the behaviour under consideration in this research. Spatial data sharing has been defined as making spatial data available to, or accessing data from, other organisations under certain terms and conditions. With regard to the target of the behaviour, spatial data, this exchange relationship entails both control over internal spatial data resources and dependence on external spatial data resources. Both need to be assessed.

The two perspectives need to be applied to the same organisation; that is, an organisation can be dependent on certain spatial data sets while it also owns data sets that, in turn, may be required by other organisations. Measuring both aspects for one organisation gives a balanced view of the organisation's perceived control over, and dependence on, spatial data.

Kevany does not define different levels of dependence. In order to measure the level, or extent, of dependence on spatial data, we draw again on the resource dependence literature. This literature suggests that organisations operate in turbulent and uncertain environments where critical resources are often controlled by other organisations (Pfeffer and Salancik, 1978; Mizruche and Galaskiewicz, 1993). The assumed goal of organisations is to ensure the flow of resources from other organisations.

This approach has been criticised for focusing mostly on the constraints of organisational designs and actions and for arguing that an organisation's selection of particular organisational forms and behaviour depends on the environmental contexts within which it operates (Knoke and Janowiec-Kurle, 1999). In their analysis of organisational networks and exchanges, Alter and Hague (1993) have addressed this point and agree that organisational activities are not entirely determined by the environment within which an organisation is functioning. They argue that the extent of an organisation's own self-sufficiency determines its dependence on its environment.

The reasons for drawing on this literature to develop appropriate proxy measures for this study are twofold. First, the resource dependence literature acknowledges the close link between the dependence on other organisations and an organisation's behaviour: 'The argument that the organization is a coalition of support implies that an important factor determining the organization's behaviour is the dependencies on the various coalition partners' (Pfeffer and Salancik, 1978: 45).

This resonates with the aim of measuring the extent to which the dependence of an organisation on external spatial data affects its willingness to engage in spatial data sharing. Second, due to the focus of resource dependence on the role of resources in inter-organisational relationships, this literature has developed a practical approach to assessing the dependence for resources of one organisation on another.

Pfeffer and Salancik (1978) argue that three factors are critical in assessing the dependence of one organisation on another: firstly, the importance of the resource, secondly, discretion over the allocation and use of the resource, and thirdly, the scope of alternative sources of the resource. These are considered below in turn together with Alter and Hague's (1993) consideration of the reliance on external resources. As in section *Attitude domain 3: Strategic position* in this chapter, the concepts developed in the resource dependence literature provide a basis on which to assess both the dependence on, and the control over, spatial data.

The concepts defined by Pfeffer and Salancik (1978) and Alter and Hague (1993) are complementary rather than exclusive. Used in combination, they can be employed to assess factors external (environmental) and internal to the organisation, thus addressing the concerns of the critics of the resource dependence approach. Considerations about the perceived control of spatial data and the perceived concentration of spatial data sources can be used to assess the external (or environmental) factors while the perceived importance of the spatial data for the organisation and the perceived reliance on external spatial data can be used as indicators to assess the factors internal to the organisation.

1. Importance of the resource
   Pfeffer and Salancik's (1978) first factor captures perceptions about how important the resource is for an organisation's functioning. Two dimensions are considered. First, the relative magnitude of the resource exchange can range from single to multiple inputs and is measurable by assessing the proportion of total inputs or outputs accounted for by the exchange. This dimension is not used here since it is more suitable for measuring actual rather than intended behaviour. The second dimension of the importance of a resource, the criticality of the resource, is the extent to which an organisation requires a resource for its continued operation and survival. For the measurement of dependence on spatial data, this can be operationalised by assessing the ability of an organisation to continue functioning in the absence of spatial data from other organisations. For the measurement of control over spatial data, the importance of the spatial data an organisation would make available to sharing partners for their continued operation can be assessed.

2. Discretion over allocation and use
   The second factor for determining the perceived dependence of an organisation is the extent of discretion over resource allocation and use. The four elements, copyright, access, use, and enforcement of rules, making up this factor have already been introduced in this chapter in section *Attitude domain 3: Strategic position*. These elements serve here to measure the control and

dependence beliefs and the perceived likelihood that each particular factor will facilitate or inhibit sharing. For the dependence of an organisation on spatial data from other organisations, the importance of these other organisations a) not copyrighting, b) not regulating access and c) use, and d) not enforcing rules regarding the ownership of, access to and use of their spatial data is assessed to measure the perceived exploratory power of these factors. The dependence beliefs are used to measure the extent to which other organisations are perceived to be able to enforce copyright, regulated access and use, and rules for their spatial data. The reverse perspective is taken when measuring the 'perceived control' of an organisation over its spatial data. The perceived importance of copyrighting its spatial data, regulating access, controlling use, and enforcing rules is assessed. The corresponding control beliefs assess the extent to which an organisation is able to ensure these control mechanisms.

3. Scope of alternative sources
    The third factor identified by Pfeffer and Salancik (1978) is the scope of alternative sources of the resource. They argue that the importance of a resource and its control by another organisation can only create dependence if the resource control is highly concentrated. This is the case when there are only a few and unstable alternative sources for the resource, implying a high reliance on one source. Applied to spatial data sharing, the dependence of an organisation on spatial data from other organisations can be measured by considering the availability of alternative sources for the spatial data it requires and the stability of these external sources. Similarly, the control an organisation has over its spatial data can be assessed by the availability of alternative sources for the spatial data that other organisations require and by the stability of these alternative sources.

4. Reliance on external resources
    Alter and Hague (1993) conceptualise the reliance of an organisation on external resources as ranging from self-sufficiency to dependence. In combination with the factors provided by Pfeffer and Salancik (1978), this provides a fourth measurement for the dependence of an organisation on external spatial data. However, it does not serve as an additional measure for the control over spatial data.

Table 5.33 Operationalisation of inter-organisational exchange concepts

Inter-organisational exchange concepts	Control over own spatial data	Dependence on others' spatial data
Importance of the resource: extent to which the organisation requires it for continued operation and survival (Pfeffer and Salancik 1978) ▪ relative magnitude of the resource ▪ criticality of the resource	Importance of the spatial data that an organisation would give to sharing partners for their continued operation.	Importance of the spatial data an organisation would receive from sharing partners for the continued operation of its organisation.
Discretion over resource allocation and use (Pfeffer and Salancik 1978): ▪ copyright ▪ access ▪ use ▪ enforcement of regulations	**Perceived importance** ▪ *Copyright:* Importance of covering the spatial data sets an organisation makes accessible to other organisations by copyright. ▪ *Access:* Importance of regulating access to an organisation's spatial data by other organisations. **Control belief** ▪ *Copyright:* Extent to which an organisation can ensure copyright on its spatial data sets that it makes available. ▪ *Access:* Extent to which an organisation can regulate access to its spatial data by other organisations.	**Perceived importance** ▪ *Copyright:* Importance of the spatial data sets an organisation requires not being covered by copyrights of other organisations. ▪ *Access:* Importance of access to the spatial data an organisation requires from other organisations not being regulated. **Dependence belief** ▪ *Copyright:* Extent to which copyright is ensured by other organisations on the spatial data sets that they make available. ▪ *Access:* Extent to which other organisations are able to regulate access to spatial data owned by them.

Inter-organisational exchange concepts	Control over own spatial data		Dependence on others' spatial data	
	Perceived importance	**Control belief**	**Perceived importance**	**Dependence belief**
Discretion over resource allocation and use continued	■ *Use*: Importance of organisations other than sharing partners not able to use spatial data owned by the organisation. ■ *Enforcement of rules*: Importance of enforcing rules regarding the ownership of, access to and use of, an org.'s spatial data.	■ *Use*: Extent to which an organisation can control the use of its spatial data by organisations other than sharing partners. ■ *Enforcement of rules*: Extent of rule enforcement by an org. regarding the ownership of, access to and use of, its spatial data.	■ *Use*: Importance of being able to use spatial data not owned by the organisation. ■ *Enforcement of rules*: Importance of other organisations not enforcing rules re. ownership of, access to and use of, their spatial data.	■ *Use*: Extent to which other organisations control the use of spatial data owned by them. ■ *Enforcement of rules*: Extent of rule enforcement by other organisations regarding the ownership of, access to and use of, their spatial data.
Extent of *few alternatives*, or the extent of control over the resource by the interest group. (Pfeffer and Salancik 1978)	■ Availability of alternative sources for the spatial data that other organisations require from an organisation. ■ Stability of the alternative sources for the spatial data that other orgs. require from an org.		■ Availability of alternative sources for the spatial data an organisation requires. ■ Stability of the external sources for the spatial data an organisation requires.	
Reliance on external resources: in inter-org. networks, the extent of reliance on ext. resources should be conceptualised as *self-sufficiency versus dependence*. (Alter and Hage 1993)			■ Self-sufficiency of an organisation in terms of spatial data resources. ■ The current position of an organisation in terms of spatial data exchanges (predominantly giving or receiving spatial data).	

A list of the beliefs about control over spatial data is given in Table 5.34 and beliefs about the dependence on spatial data are shown in Table 5.35. A summary of how these are operationalised is presented in Table 5.33. A specific example of the corresponding questionnaire items for these beliefs is given in Table 5.36.

Table 5.34   Control beliefs about control over spatial data

Belief
1. Importance of spatial data to sharing partners
2. Copyright
3. Regulated access
4. Use
5. Rule enforcement
6. Availability of alternative sources
7. Stability of alternative sources

Table 5.35   Control beliefs about dependence on spatial data

Belief
1. Importance of spatial data to own organisation
2. Copyright
3. Regulated access
4. Use
5. Rule enforcement
6. Availability of alternative sources
7. Stability of alternative sources
8. Spatial data self-sufficiency of the organisation
9. Spatial data sharing position

Table 5.36   Sample questionnaire items: belief about spatial data position

a) Control belief							
The extent to which your organisation could regulate access to its spatial data by other organisations in some way other than ownership is likely to be...:							
1	2	3	4	5	6	7	□
(not at all) extrem. small	slightly	moder- ately	fairly	quite	very	extrem. large	no op./ not relev.
b) Perceived importance							
How important is it that access to your organisation's spatial data by other organisations be regulated in some way other than ownership?							
1	2	3	4	5	6	7	□
not at all important	slightly	moder- ately	fairly	quite	very	extrem. important	no op./ not relev.

In summary, in the previous four domains (technical skills, interpersonal skills, control over resources and spatial data position), internal control factors such as skills and resources that may be considered necessary by key individuals within organisations to engage in spatial data sharing across organisational boundaries have been dealt with.

External Factors

Under 'perceived control', internal resources central to spatial data sharing are considered and the extent to which they are perceived to be under the organisations' control. Similarly, 'perceived control' also includes external factors of dependence, such as people or resources outside the organisation, upon which organisations may perceive to be dependent. The external factors are organised into two domains: finding sharing partners, and sharing opportunities.

Perceived control domain 5: dependence on others - finding sharing partners The fifth domain that is likely to affect 'perceived control' is finding sharing partners. One of the external factors is the dependence on the co-operation of other people (Ajzen and Madden, 1986). According to Ajzen (1985), the higher the perceived dependence on the co-operation of other people, the lower the perception of an individual's own control over the behaviour. The dependence on others is seen as an impediment to performing the behaviour that can lead to temporary changes in intentions: 'Whenever performance of a behaviour depends on the actions of other people, there exists the potential of incomplete control over behavioural goals' (Ajzen, 1985: 28).

For sharing spatial data, the dependence on other people or organisations is created by having to find sharing partners with whom to share spatial data. The ease or difficulty of finding sharing partners can be assessed in terms of the perceived ease or difficulty of finding organisations with the appropriate characteristics to qualify them as sharing partners. There are four characteristics of potential sharing partners that can be considered (see Table 5.37).

Table 5.37   Control beliefs about sharing partners

Belief
1.   Willing sharing partners
2.   Reliable sharing partners
3.   Compatible purpose of application
4.   Organisational fit

First, the *willingness of sharing partners* is deemed relevant because requests for spatial data can be met with refusal or unacceptable conditions (Craig, 1995) or the owner of data may offer access to it but then refuse to share it (Kevany, 1995). The willingness of sharing partners has been discussed by Kevany (1995) as follows:

> The owner of the digital data or the sources for data must be willing to share the data
> with other organizations ... The organizations that want to gain access to the data of

another or to share the development of data must be willing to commit the resources necessary for sharing. That may include fees to the owner of data or source materials or the cost of compilation or conversion of the data to be shared. (Kevany, 1995: 83)

The *reliability of sharing partners* means primarily that they honour copyright or other distribution restrictions. It also means that once agreements are made they are not broken later on (Tosta, 1995).

Regarding the *purpose* for which shared spatial data is acquired, interviewees in Phase I indicated that in South Africa there is uncertainty stemming from the Apartheid era about the use of shared spatial data and a perceived lack of control over who uses the information and for what purpose.[31] Therefore, another characteristic of the potential sharing partner to be considered is a compatible purpose of application of the spatial data.

Finally, with respect to the type of organisation of the sharing partner, Bamberger (1995) suggests that individuals in many organisations often do not understand or trust the motives of individuals in other organisations. An *'organisational fit'*, then, can be defined as the type of organisation (of the sharing partner) one is comfortable with. An example of the corresponding questionnaire items for these beliefs is provided in Table 5.38.

Table 5.38   Sample questionnaire items: belief about finding sharing partners

a) Control belief							
For our organisation, finding sharing partners with the following characteristics would be...: 1. Reliable							
-3	-2	-1	0	1	2	3	☐
extrem. difficult	quite	slightly	uncertain	slightly	quite	extrem. easy	no op./ not relev.
b) Perceived importance							
How important are the following characteristics of any current or prospective spatial data sharing partners? 1. Reliability							
1	2	3	4	5	6	7	☐
not at all important	slightly	moder- ately	fairly	quite	very	extrem. important	no op./ not r.

---

[31] Similarly, Onsrud (1995) suggested that sharing partners may need to be made aware that no guarantees, implied or otherwise, can be given about the suitability of the spatial data for the purpose intended by the receiver.

Perceived control domain 6: sharing opportunities   The last domain underlying 'perceived control' over sharing spatial data is sharing opportunities. Ajzen (1985) discussed the role of circumstantial factors or opportunities. In contrast to unanticipated events which can be associated with changes in intention, the absence of opportunities to perform a particular behaviour is not expected to affect intention and its underlying determinants. By implication, the presence of opportunities can facilitate the performance of the behaviour.

In the South African context, the NSIF has been conceptualised in line with the definition of the US National Spatial Data Infrastructure (NSDI): '... the technology, policies, standards and human resources necessary to acquire, process, store, distribute and improve the utilization of geospatial data' (Federal Register, 1994: Section 1(a)).

A directorate has been set up within the Department of Land Affairs which was endorsed by the Cabinet (see Chapter 2). Owing to the intended role of the NSIF directorate as a facilitator for spatial data sharing, it is considered here within the opportunity creation domain. In order to assess the extent of its opportunity creation for spatial data sharing, several specific activities of the NSIF are considered (see Table 5.39).

Table 5.39   Control beliefs about sharing opportunities

Belief
1. NSIF awareness creation
2. Fora organised by the NSIF
3. Policy development
4. Standards alignment
5. South African core data set identification
6. Spatial Data Discovery Facility

First, NSIF awareness creation activities refer to the promotion of the benefits of co-operation and sharing (although primarily among public sector agencies) and the provision of guidance on general sharing issues.

Second, the importance of opportunities to meet with other members of the GIS community was a frequently mentioned belief during the pilot interviews in Phase I and it has also long been recognised in the literature. Granovetter (1972) has pointed out that the maintenance of weak ties of networks may be one of the most important aspects of meetings. The provision of a communication structure and a forum for debate is one of the objectives of the NSIF (NSIF, 1998a). Hence the fora organised by the NSIF provide a specific point of reference for assessing the importance of meeting opportunities.

Next, policies, standards, core data sets and clearinghouses are central elements of spatial data initiatives such as the NSIF (see Chapter 2). The clearinghouse of the NSIF, the Spatial Data Discovery Facility, is an on-line data catalogue containing metadata with Internet links to distributed sites and tools to exploit the catalogue and find information (NSIF, 1998a). Perceptions about furthering of the development and implementation of policies and standards, a South African core data set and the Spatial Data Discovery Facility are additional items that can be used to assess the scope and relevance of the NSIF opportunity creation. An example of the corresponding questionnaire items for these beliefs is provided in Table 5.40.

Table 5.40　Sample questionnaire items: belief about sharing opportunities

b) Perceived importance							
How important would be the following activities by the National Spatial Information Framework (NSIF) for your organisation's spatial data sharing activities? 1. NSIF awareness creation							
1	2	3	4	5	6	7	☐
not at all important	slightly	moder-ately	fairly	quite	very	extrem. important	no op./ not relev.

Summary of Perceived Control Determinants

In this section, the factors that may influence one of the main elements of the larger model, 'perceived control', have been discussed. The domains underlying the 'perceived control' over spatial data sharing are presented in Figure 5.5.

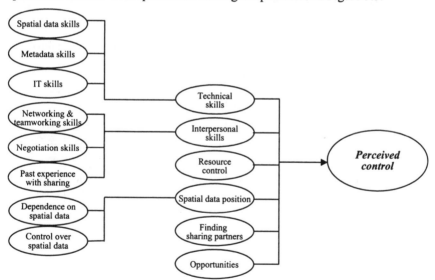

Figure 5.5　Determinants of 'perceived control' over spatial data sharing

Specifically, the domains that are expected to influence 'perceived control' over spatial data sharing are: 1) technical; 2) interpersonal skills; 3) control over resources; 4) spatial data position; 5) finding sharing partners; and 6) opportunities for sharing. The beliefs compiled in these domains in line with the TPB discussed in Chapter 3 are expected to influence an organisation's 'perceived control' over spatial data sharing.

An example of the questionnaire items at the level of the domains of the 'perceived control' component is provided in Table 5.41.

Table 5.41   Sample questionnaire items: perceived control domain-level (technical skills)

For spatial data sharing, our organisation is likely to have the necessary technical capabilities (in spatial data, Metadata, and computing). 1. at the moment							
-3	-2	-1	0	1	2	3	☐
strongly disagree	quite	slightly	uncertain	slightly	quite	strongly agree	no op./ not relev.
2. in the next month – 2 years							
-3	-2	-1	0	1	2	3	☐
strongly disagree	quite	slightly	uncertain	slightly	quite	strongly agree	no op./ not relev.

Table 5.42 presents an example of the two 'direct' measures of the 'perceived control' over spatial data sharing.

Table 5.42   Sample questionnaire item: direct measure of perceived control

It is mainly up to us whether our organisation engages in spatial data sharing activities. 1. at the moment							
-3	-2	-1	0	1	2	3	☐
strongly disagree	quite	slightly	uncertain	slightly	quite	strongly agree	no op./ not relev.
2. in the next month – 2 years							
-3	-2	-1	0	1	2	3	☐
strongly disagree	quite	slightly	uncertain	slightly	quite	strongly agree	no op./ not relev.

Summary: A Model of the Willingness to Share Spatial Data in South Africa

By extending the TPB, a range of likely determinants of the willingness of organisations to engage in spatial data sharing with other organisations has been identified above. Together, they provide the components of a model of the perceived willingness to engage in spatial data sharing as shown in Figure 5.6.

The structure of the model (from *right* to *left*) is as follows: spatial data sharing behaviour is determined by the intention, or the willingness, to engage in spatial data sharing. The willingness to share spatial data across organisational boundaries is assumed to arise from the 'attitude' towards spatial data sharing, 'social pressure' to share and 'perceived control' over spatial data sharing activities. These three components, shown at the centre of Figure 5.6, are assumed to be formed on the basis of domains (or groups) of underlying beliefs. The domains are presented on the left of the diagram.

Specifically, the 'attitude' towards spatial data sharing is suggested to be formed on the basis of beliefs about the possible positive or negative outcomes that may result from spatial data sharing activities. The 'social pressure' with respect to spatial data sharing is suggested to arise from beliefs of key individuals about the expectations from the social context within which their organisations operate. The 'perceived control' over spatial data sharing activities concerns beliefs about the extent of control over the necessary skills, resources and opportunities for spatial data sharing activities.

This model can be used in an applied setting by operationalising its elements in a questionnaire instrument to understand the willingness of organisations, as perceived by their key individuals, to engage in spatial data sharing across organisational boundaries. In line with the method and research design, the model provides the basis for Phase II, the quantitative empirical stage, consisting of a survey of spatial data sharing perspectives in South Africa. The methods for compiling the questionnaire based on this model have been discussed in Chapter 4. The model is used to derive hypotheses in order to test the relationships between the elements of the model against the empirical data collected in Phase II.

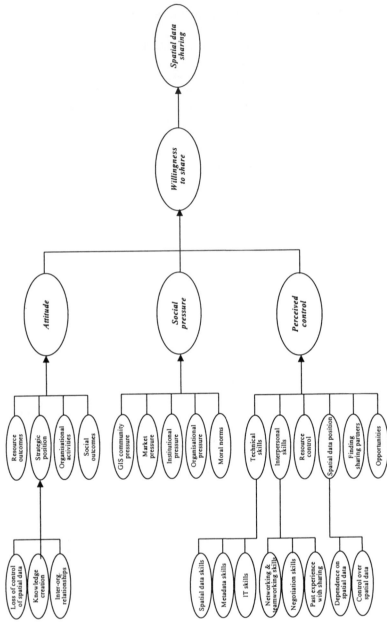

Figure 5.6 Model of the willingness to share spatial data across organisational boundaries (South Africa)

Hypotheses

The model of the likely determinants of the willingness to share spatial data presented in the previous section can be used to generate a number of hypotheses regarding the likely relationships between its constructs or elements by considering the model in the light of the TPB.

Relational Hypothesis

The TPB suggests that beliefs form the basis for the formation of three main constructs – 'attitude', 'social pressure', and 'perceived control'. The following structure of the likely relationship between the direct and the belief-based measures was proposed by Ajzen (1991). These relationships are hypothesised for the model of the perceived willingness of organisations to engage in spatial data sharing.

Attitude - behavioural beliefs relationship According to the TPB, the positive or negative evaluation (e) of each outcome associated with spatial data sharing is multiplied with the strength (i.e. likelihood) of these associations (b) to provide a measure of the behavioural belief (be). As stated in equation one, it is hypothesised that the 'attitude' towards spatial data sharing (A) is related to the sum of behavioural beliefs (Σbe) about outcomes from spatial data sharing.

Hypothesis 1:
$$A \propto \sum_i^n b_i e_i \tag{1}$$

Perceived social pressure - normative beliefs relationship Each normative belief (nm) is considered to consist of the strength of the normative belief (n) (indicated by the perceived approval or disapproval by important referents of the engagement in spatial data sharing) which is multiplied with the motivation to comply (m) with the referent in question. 'Social pressure' to engage or not to engage in spatial data sharing is hypothesised to be related to the sum of normative beliefs (Σnm) regarding the expectations of important referents, as stated in equation two.

Hypothesis 2:
$$PSP \propto \sum_i^n n_i m_i \tag{2}$$

Perceived control - control beliefs relationship Control beliefs about resources (cp) are expected to consist of the perceived importance of each skill or resource (p) which is multiplied by the 'perceived control' over the skill or resource (c). As formally expressed in equation three, 'perceived control' over spatial data sharing is hypothesised to be related to the sum of control beliefs (Σcp) about the availability of the skills and resources for sharing spatial data.

Hypothesis 3:
$$PBC \propto \sum_i^n c_i p_i \tag{3}$$

Explanatory Hypotheses

The second set of hypotheses derives from the explanatory power of the TPB. This is expected to apply at two levels.

Explanatory level one   At level one, the three main components - 'attitude' (*A*), 'social pressure' (*PSP*) and 'perceived control' (*PBC*) - are hypothesised to explain the perceived willingness to share spatial data (*W*), as stated formally in equation four.

Hypothesis 4: $W = W (A, PSP, PBC)$ (4)

Explanatory level two   At level two, beliefs are assumed to provide the basis for 'attitude', 'social pressure' and 'perceived control'. Therefore, measures of the beliefs should serve to explain each of those components.

Under *'attitude'*, each of the four domains (resource outcomes, strategic position, organisational activities, and social outcomes) consists of one or more scales.[32] The combined beliefs (*be*) belonging to a particular scale are summed to form a scale index (Σbe). Hence, 'attitude' (*A*) is hypothesised to be a function of the resultant scale indices (see equation five).

Hypothesis 5: $A = A$ (Σcosts, Σbenefits,   (5)
                       Σspatial data outcomes,
                       Σorganisational activities,
                       Σloss of control, Σknowledge creation,
                       Σinter-org relations, Σsocial outcomes)

'Perceived social pressure' consists of five domains (pressure from GIS community, and market, institutional, and organisational pressure, and moral norms). The combined beliefs (*nm*) belonging to a particular domain are summed to form a scale index (Σnm). 'Social pressure' (*PSP*) is hypothesised to be a function of the resultant scale indices, as indicated in equation six.

Hypothesis 6: $PSP = PSP$ (ΣGIS community, Σmarket,   (6)
                       Σinstitutional, Σorganisational
                       pressure, Σmoral norms)

Each of the six domains under *'perceived control'* (technical skills, interpersonal skills, resource control, spatial data position, finding sharing partners, and opportunities) consists of one or more scales. The combined beliefs (*cp*) belonging to a particular scale are summed to form a scale index (Σcp). Therefore,

---

[32] A scale is made up of a number of related items to measure a single concept (Bryman and Cramer, 1999).

'perceived control' (*PBC*) is hypothesised to be a function of the resultant scale indices (see equation seven).

Hypothesis 7:

$$PBC = PBC \ (\Sigma \text{spatial data}, \Sigma \text{metadata}, \quad (7)$$
$$\Sigma \text{itskills}, \Sigma \text{networking},$$
$$\Sigma \text{negotiation}, \Sigma \text{past experience},$$
$$\Sigma \text{resource control},$$
$$\Sigma \text{importance of internal data},$$
$$\Sigma \text{control aspects}, \Sigma \text{altern. sources1},$$
$$\Sigma \text{importance of external data},$$
$$\Sigma \text{dependence aspects},$$
$$\Sigma \text{altern. sources2}, \Sigma \text{self sufficiency},$$
$$\Sigma \text{sharing partners}, \Sigma \text{opportunities})$$

The extent to which these relationships hold in the model of the perceived willingness of organisations to engage in spatial data sharing is considered in Chapter 7 where the model is validated against the empirical survey data. The reasoning and procedure for deriving these hypotheses was outlined in the methodology in Chapter 4. Substantiation of the relationships that are proposed here to exist between the elements of the model is a necessary step so that the empirical data resulting from the application of the model can be relied on to address the research questions.

Past Sharing Behaviour

In addition to measuring the determinants influencing behaviour, the TPB also provides constructs for measuring the behaviour in question. These include the categorisation of a single action versus behavioural categories and the definition of the behavioural elements (action, target, context, and time) (Ajzen and Fishbein 1980). This research employs the TPB with respect to its ability to *explain* an intention to perform a behaviour, measuring the underlying beliefs that are expected to influence 'attitude', 'social pressure' and 'perceived control' in order to understand why individuals within organisations may, or may not, appear to want to engage in spatial data sharing. In cases like this, where the theory's ability to *predict* behaviour from intention is not of central importance, the relationship between intention and behaviour is not of primary concern. Hence a complete assessment of the behaviour, spatial data sharing, through observing and recording it subsequent to administering the questionnaire is not necessary (Ajzen 1985). Yet in order to respond to the second research question regarding the extent of actual spatial data sharing activities, some assessment of actual sharing behaviour has to be included.

An alternative means of examining actual behaviour to direct observation of that behaviour is by using self-reports of past behaviour. Ajzen and Fishbein (1980) suggest that, in many instances, the observation of the behaviour by an outside observer presents considerable difficulties. As this is the case with spatial data sharing across organisational boundaries and since this method would also encompass a considerable effort that is beyond the scope of this research, self-reporting of past sharing behaviour was chosen. Ajzen (1988) has argued that this is an acceptable method for relatively non-sensitive behaviours in which the relationship between people's self-reports and their actual behaviour is high. In this respect, spatial data sharing can be regarded as non-sensitive and respondents to the survey in Phase II were asked to report on their past sharing behaviour.

The approach proposed here to measure past spatial data sharing behaviour is based on Calkins and Weatherbe's (1995) taxonomy for spatial data sharing by and is presented in Table 5.43. Multiple choice questions for the measurement of sharing behaviour have been devised (see Appendix B). This is based on the understanding that a comprehensive list of activities that may be considered sharing activities would constitute a questionnaire in its own right.

Table 5.43   Measurements of past sharing behaviour

Aspects of Past Sharing Behaviour	
Type of the organisation	private sector / commercial
	academic institution
	non-governmental organisation
	local authority
	provincial government
	national government
	para-statal organisation
Spatial data role of the organisation	spatial data recipient
	spatial data supplier
	spatial data broker
Type of spatial data exchange of the organisation	data supplier <-> end user
	data suppliers <-> data broker
	end user <-> data broker
	between data suppliers
	between end users
	between data broker
Types of sharing partners of the organisation	vendor
	private sector / commercial
	academic institution
	non-governmental institution
	local authority
	provincial government
	national government
	para-statal organisation
Schedule of spatial data exchanges	on schedule (regularly)
	on demand (ad hoc)

Table 5.43   continued

Aspects of Past Sharing Behaviour	
Frequency of spatial data exchanges	daily
	weekly
	monthly
	yearly
	project-basis
	once-off
	never
Most recent spatial data exchanges	this week
	last week
	last month
	last 6 months
	more than 6 months ago
Sharing arrangement	informal / voluntary
	formal contract
	mandate to share or distribute
	profit-making venture
Charges for shared spatial data	free of charge
	barter
	license fee
	consortium membership
	marginal cost of reproduction
	cost recovery basis
	market value
Average quantity	single observation
	summary / aggregate observations
	selected subset
	theme
	entire database

Source: based on Calkins and Weatherbe (1995)

The following information was gathered. The type of the organisation, the spatial data role and exchanges of the organisation, and the type of sharing partners provide information on who is sharing with whom. The average quantity of spatial data shared, together with the schedule, frequency and most recent spatial data exchanges supply information about the magnitude of spatial data exchanges. Information about the sharing arrangements and the charges for shared spatial data are used to give insight into the terms and conditions under which spatial data are shared.

Conclusion

This chapter has combined interview material from Phase I with insights drawn from several bodies of literature to develop a model of the willingness of individuals in organisations to engage in spatial data sharing across organisational boundaries. Based on the TPB as an organising framework, proxy measures of the likely determinants of intended spatial data sharing behaviour were identified which, together, comprise the components of the model. Several hypotheses were derived regarding the likely relationships between the constructs included in the model and regarding the explanatory power of the model. In addition, measures were also devised for examining the past sharing behaviour of organisations.

The next chapter presents the results of Phase II of the empirical research which consisted of implementing the model presented here in a questionnaire-based survey in South Africa. These results provide first insights into the incentives and disincentives for spatial data sharing across organisational boundaries and they provide the basis for addressing the first and the second research question: how willing are key individuals embedded in organisations to engage in spatial data sharing and what is the extent of actual sharing activities in South Africa?

Chapter 6

Survey of Spatial Data Sharing Perspectives

Introduction

The aim of this study is to further the understanding of how spatial data sharing can be fostered so that bottlenecks in the availability of, and access to, spatial data may be overcome. To this end, the study seeks to identify the existing incentives and disincentives to participating in spatial data sharing that may be present within the community of actors. These insights are expected to provide a basis for addressing the issue of spatial data sharing more effectively at the policy making level.

This chapter presents the results from the *quantitative* empirical stage of the research design that was described in Chapter 4. Phase II of the empirical research consisted of a survey of the spatial data sharing perspectives of key individuals within organisations that are involved with GIS applications. Based on the model of the perceived willingness to share spatial data that was presented in Chapter 5, a questionnaire was developed and administered in South Africa over a two-month period. The procedures were described in Chapter 4. Descriptive statistics are used in this chapter to report the distribution of the scores for the variables included in the questionnaire items and to provide a first insight into the determinants of spatial data sharing behaviour.

Following a brief discussion of the scales and their interpretation, the results are presented following the structure of the model of the perceived willingness to share spatial data that was introduced in Chapter 5. Hence, first an assessment is presented of the apparent overall willingness of the sample respondents to engage in spatial data sharing. Next, the results are presented for 'attitude' and its four underlying domains; for 'social pressure' and its five domains; and for 'perceived control' and its six domains. Finally, the self-reports of actual spatial data sharing in South Africa are summarised.

Explanation of the Measurements and their Interpretation

The variables included in the model of the perceived willingness to engage in spatial data sharing were measured using the questionnaire items that incorporated differential semantic scales. Semantic differentials provide a means of capturing the connotative meaning of the concept being assessed by including all of its suggested or implicit significance (Aronson *et al.*, 1990). Implemented on a seven point unipolar or bipolar scale, the marks placed closest to the ends of the scales show that the adjective at the end of the scale strongly reflects the respondent's opinion with respect to the concept that is addressed by the questionnaire item. Scores at the centre of the scale indicate a respondent's ambivalence or indecision.

In the presentation of the results of the survey in this chapter, the distribution of the responses to each question is indicated by the percentage of respondents who chose each point on the scale. For each *end* point of the scale, the two scores placed closest to it were jointly interpreted as representing those respondents who were clearly expressing their beliefs. Those respondents choosing one of the three mid points of a scale were classed as being undecided.

Each of the questionnaire items at the level of the main components ('attitude', 'social pressure', 'perceived control' and intention) and at the level of the domains related to two distinct time frames (the present and the near future), adding a second dimension to each question.[33]

Assessment of the Willingness to Share Spatial Data

The starting point of this research was the recognition that current spatial data infrastructure initiatives take the willingness of various organisations to engage in spatial data sharing activities more or less for granted. The use of a behavioural approach as an organising framework for this study has enabled a detailed assessment of the willingness of the organisations to share, as perceived by the respondents in the survey, in order to address the first research question: How willing are key individuals embedded in different organisations to engage in spatial data sharing across organisational boundaries?

The willingness to share spatial data across organisational boundaries was measured using the intention construct, with reference to the present and to the near future (see Chapter 3 for an explanation of this construct).[34] The conceptual location of the intention construct within the model of the willingness to engage in spatial data sharing is highlighted in Figure 6.1.

---

[33] Near future refers to the time period 'in the next month to two years'.

[34] For the intention construct, three items were included in the questionnaire to ensure that at least one measure of this important component of the model would be usable. A preliminary set of regression analyses was conducted to determine the single intention item whose variance could be best explained by the three components of the model ('attitude', 'social pressure' and 'perceived control'). The measure for intention included in subsequent analyses of the data was the item '*Considering the outcomes, pressures, capabilities and resources, the likelihood for your organisation to engage in spatial data sharing is low/high*'.

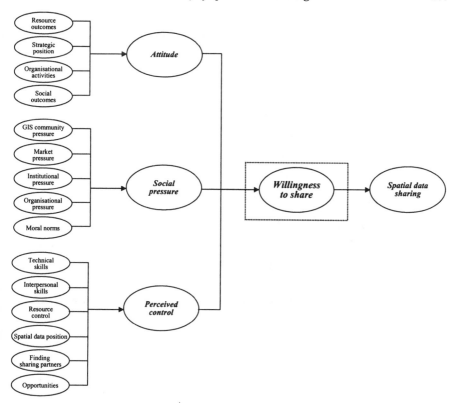

Figure 6.1 **Model of the willingness to share spatial data: focus on intention**

The mean scores of this variable are presented in Table 6.1. They indicate a slight increase in the willingness to share spatial data in the near future compared to the present.

Table 6.1 **Direct measures of intention**

	Current		Near future	
Direct measures	**Mean**	**Standard deviation**	**Mean**	**Standard deviation**
Intention	4.58	1.50	5.56	1.38

Note: Scale ranged from '1 - extremely low' to '7 - extremely high' likelihood for the organisation to engage in spatial data sharing.

(n = 112)

The distribution of the scores for the measure of intention (at present) is presented in Figure 6.2. The apparent willingness to share spatial data across organisational boundaries is indicated by the respondents choosing the extreme values. Hence, the two scores at the positive end of the scale indicate an apparent willingness to share while the two scores at the other extreme indicate resistance to share. The middle scores indicate that respondents were undecided either way.

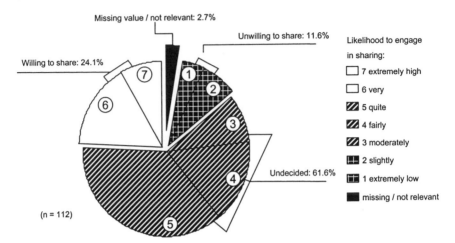

Figure 6.2   Assessment of the willingness of organisations in South Africa to engage in spatial data sharing

As can be seen in Figure 6.2, the finding of this research is that the actual motivation of organisations in South Africa was not in favour of participating in spatial data sharing. Overall, the willingness to share spatial data across organisational boundaries was low in the sample; only a quarter of the respondents reported that their organisation was very willing to share. Nearly 62% of the respondents reported that their organisation was uncertain as to whether to engage in spatial data sharing with other organisations. A further 12% indicated that their organisations were unwilling to share spatial data across organisational boundaries.

The willingness to share among the sample respondents was analysed to examine whether there are any significant differences in these intentions based on three criteria: 1) differences in intention between the different sectors of the GIS community, 2) differences in intention depending on the organisation's reliance on external spatial data, and 3) differences in intention depending on the organisation's spatial data position in spatial data exchanges. The detailed results of this analysis are presented in Appendix C.

The results of the analysis indicate that there was no statistically significant difference in the willingness expressed by the respondents representing different sectors of the GIS community. Similarly, the perceived willingness of respondents regarding their organisation's engagement in spatial data sharing did not differ according to their organisation's reliance on external spatial data resources, i.e. whether respondents ranked their organisation as self-sufficient or dependent. Finally, there was also no statistically significant difference between the willingness to share according to how respondents perceived their organisation's spatial data position in spatial data exchanges, i.e. whether they ranked their organisation as predominantly giving versus predominantly receiving spatial data.

It can be concluded that the willingness of organisations, as expressed by the respondents in the sample, to engage in spatial data sharing across organisational boundaries did not vary for the different sectors in the GIS community, nor did it differ with the extent of perceived self-sufficiency or dependence of organisations on spatial data or the organisation's spatial data position in terms of predominantly giving spatial data to, or receiving spatial data from, other organisations.

Attitude - Expected Outcomes from Sharing Spatial Data

In order to assess the 'attitude' towards spatial data sharing across organisational boundaries, the questionnaire included both a direct measure of 'attitude' and each of its domains, and indirect measures of 'attitude' by assessing the underlying beliefs within the four domains of 'attitude'.[35] The conceptual location of the 'attitude' component and its underlying domains is highlighted in Figure 6.3.

The direct measure for the 'attitude' towards spatial data sharing was assessed with reference to the present and to the near future. The central tendency, indicated by the mean, of the scores for this variable is presented in Table 6.2. The mean score for the *current* 'attitude' variable shows that the 'attitude' towards sharing spatial data across organisational boundaries among the respondents in the sample was only slightly positive. The mean score for the *near future* 'attitude' measure indicates a slight increase towards a more positive 'attitude' towards sharing. Similarly, the mean scores for the direct measures at the domain level show that the outcomes of sharing were perceived to be only slightly positive at present and somewhat more positive in the near future.

The results of assessing the behavioural beliefs that are expected to form the basis for the 'attitude' towards spatial data sharing are presented for each of the four domains. Behavioural beliefs were assessed by asking respondents to *evaluate* whether specific consequences of spatial data sharing would be positive or negative and to indicate the *likelihood* of these consequences.

---

[35] In the context of this research, the term '*direct*' measure refers to questions at explanatory level 1 in order to distinguish these questions about 'attitude', 'social pressure' and 'perceived control' from the '*indirect*' questions about 'attitude', 'social pressure' and 'perceived control' at the level of the beliefs. It is acknowledged, however, that, from an epistemological point of view, these measures can be considered only *relatively* direct measures of 'attitude', 'social pressure', and 'perceived control'.

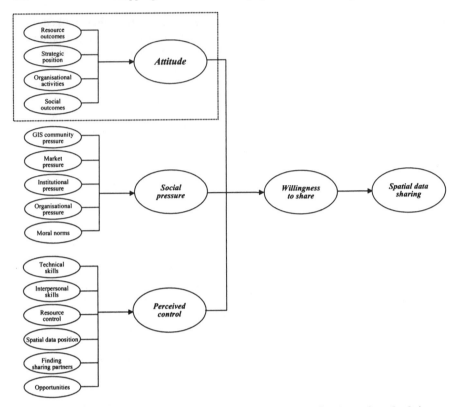

Figure 6.3   Model of spatial data sharing: focus on attitude and underlying domains

Table 6.2   Direct measures of attitude and underlying domains

	Current		Near future	
Direct measures	**Mean**	**Standard deviation**	**Mean**	**Standard deviation**
Attitude	4.92	1.09	5.66	1.06
Domain-level				
Resource outcomes	4.90	1.15	5.69	1.14
Organisational activities outcomes	5.13	.98	5.85	.92
Strategic position outcomes	4.97	1.36	5.55	1.52
Social outcomes	4.53	1.54	5.31	1.48

Note: For the sake of clarity, responses have been rescaled to '1 - extremely negative' to
        '7 - extremely positive' outcomes.
        (n = 112)

Perceived Resource-related Outcomes

The first 'attitude' domain is made up of beliefs about resource outcomes. A number of resources can be affected by sharing spatial data across organisational boundaries (see Chapter 5). The respondents evaluation of the outcome for each resource and the results with respect to the expected costs and benefits (or savings) are presented in Figure 6.4.

Significant cost implications were perceived by less than 15% of the respondents (indicated by 'extremely' and 'very significant' scores) in terms of the time required to locate spatial data outside their own organisation, requiring staff to exchange and integrate spatial data, requiring data storage for spatial data to be shared, and incurring administrative efforts and expenses from sharing.[36]

Matching the perception of low costs, significant benefits resulting from sharing were perceived by many respondents. In particular, the time saved by not having to capture spatial data in-house was indicated to be a significant saving by 62% of the respondents. More than 50% expected significant savings from sharing by freeing up trained staff who would otherwise be needed for capturing and maintaining spatial data and from significant expense savings in terms of time, equipment, data storage and labour. Significant savings in terms of data storage and administrative efforts were expected only by 33% and 37% of the respondents respectively.

---

[36] In this and in all subsequent figures in this chapter, the extreme scores of the data (i.e., or 1 and 2, or 6 and 7) are used to calculate the percentages that are given in the text. Hence, each percentage refers to one or more scores on the variables. In order to make the text flow, the full details of each column are not repeated.

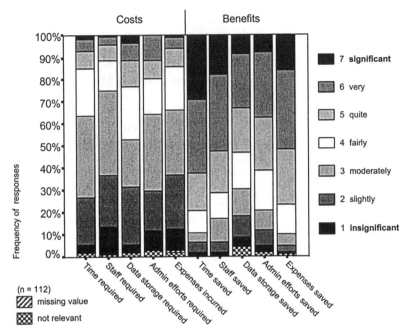

Figure 6.4   Evaluation of resource-related outcomes of sharing

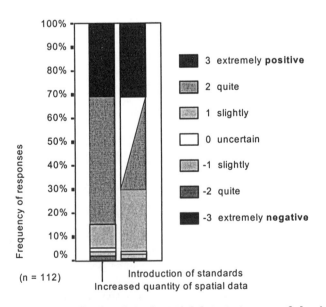

Figure 6.5   Evaluation of spatial data outcomes of sharing

Regarding the possible outcomes of sharing for spatial data, an increase in the quantity of spatial data available to their organisation was considered to be extremely positive by 85% of the respondents (see Figure 6.5). Similarly, 70% responded that the introduction of standards that are agreed in collaboration with other organisations would be extremely positive.

The respondents also assessed the likelihood of the resource-related consequences (see Figure 6.6).

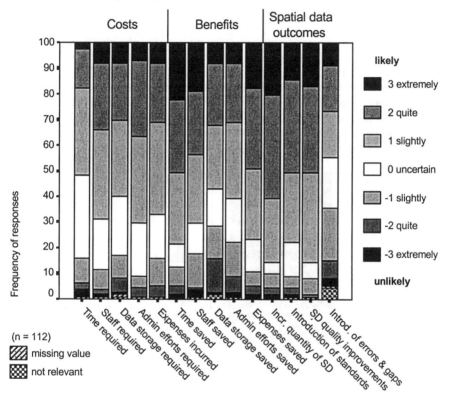

Figure 6.6   Perceived likelihood of resource-related outcomes

About one third of the respondents expected that the cost implications of spatial data sharing were highly likely, with the exception that less than 20% considered it likely that sharing would require time to locate spatial data sources outside their own organisation. The benefits considered most likely were time and cost savings (by about 50% of the respondents). Between 30-45% of the respondents indicated that savings in terms of staff, data storage and administrative efforts were highly likely to follow from sharing spatial data with other organisations. About 60% thought it highly likely that spatial data sharing would increase the quantity of spatial data available to their organisation and just above 50% considered that sharing would result in the introduction of spatial and attribute data standards and in quality improvements for the shared spatial data sets. In contrast, the introduction of errors and gaps in the shared data sets was regarded highly likely by less than 30% of the respondents.

In summary, in terms of their organisations' resources, there were more respondents perceiving benefits to arise from spatial data sharing activities than those expecting costs.

Perceived Outcomes for Organisational Activities

The second domain under 'attitude' deals with beliefs about organisational activities. Three aspects of organisational activities that may be affected by spatial data sharing activities were evaluated (see Figure 6.7). Sharing was perceived by 50% of the respondents to have strong positive consequences for their organisation's core activity. The usefulness of their organisation's GIS was expected to be positively affected by sharing by just over 80% of respondents. Finally, the effects of sharing on the quality of decision-making within their organisation were regarded to be positive by 75% of the respondents.

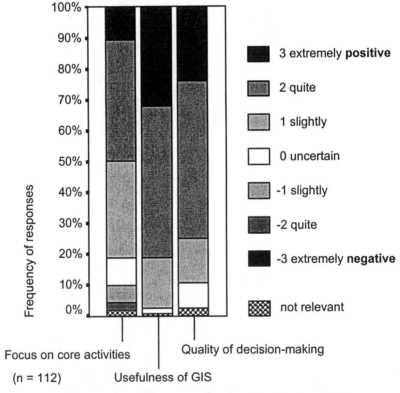

Figure 6.7   Evaluation of outcomes for organisational activities

Overall, the consequences of spatial data sharing activities for their organisation's qualitative work were perceived to be very positive by the majority of the respondents.

Perceived Outcomes for Strategic Position

The strategic position domain is the third domain under 'attitude'. It is divided into three groups of beliefs; loss of control over spatial data, knowledge creation and inter-organisational relations (see Chapter 5). The results for each area are presented in turn.

Perceived loss of control over spatial data Four measures indicate the impact of perceived loss of control over spatial data on an organisation's strategic position. The share of respondents perceiving various aspects of losing control over spatial data as a threat was generally larger than those perceiving them as opportunities (see Figure 6.8a). Concerns were indicated about unregulated access to an organisation's spatial data and lack of enforcement of rules and regulations. Nearly 40% of the respondents considered this a threat compared to 10-16% who regarded it as an opportunity. Not enforcing copyright on shared data and the use of spatial data by organisations other than the sharing partners was reported as presenting a threat by about 30% of respondents compared to just over 20% who considered this an opportunity.

Regarding the likelihood that sharing would result in loss of control over spatial data (see Figure 6.8b), almost 40% of respondents indicated it would be *unlikely* that their organisation's spatial data would not be copyrighted and that access to their spatial data would not be unregulated. On the other hand, about 25% of the respondents thought it *highly likely* that their organisation would not copyright or regulate access to their spatial data. The likelihood that organisations other than the sharing partners would be able to use their own organisation's spatial data was deemed high by 42% of respondents. Almost 30% also thought it highly likely that they could not enforce rules regarding the ownership of, access to, and use of, their organisation's spatial data.

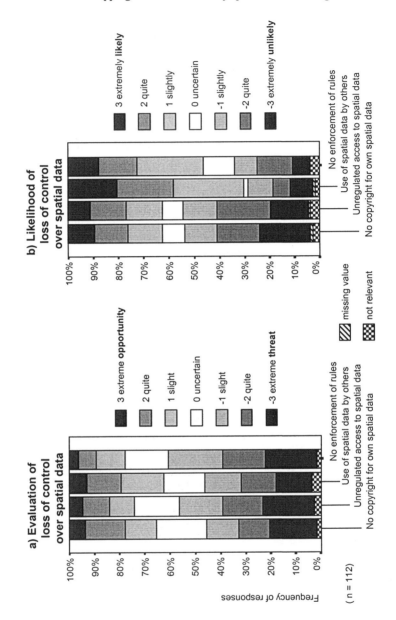

Figure 6.8   Loss of control over spatial data

Knowledge creation   In the second group of strategic position beliefs, the consequences of sharing in terms of knowledge creation and transfer were evaluated very positively (see Figure 6.9a).

The combination of new and existing spatial data sets from different organisations to create new knowledge was considered an opportunity by 75% of the respondents. Allowing access to ideas that are codified in spatial data and their translation, combination, and application in different organisations as a result of sharing was seen as an opportunity by nearly 65%. Providing the means to observe spatial data skills in personal interaction and to create shared mental models and experiences across organisational boundaries presented an opportunity for just over 70% of respondents. Finally, 76% thought it opportune if sharing were to trigger dialogue and collective reflection across organisational boundaries and result in articulating, making explicit and comparing spatial data-related concepts. These outcomes were considered very likely by many (56-74%) of respondents (see Figure 6.9b). These results show that out of the four knowledge creation 'modes' defined by Nonaka (1994) and Nonaka et al. (1996), two specific ones were evaluated more positively and as being more likely by the respondents: first, the creation of new explicit knowledge through the combination of spatial data and second, the externalisation of knowledge by triggering dialogue across organisational boundaries. This suggests that, in terms of Polanyi's (1966) distinction between tacit and explicit knowledge, the creation of explicit knowledge as an outcome of their organisations' sharing activities was valued by more respondents in the sample than the creation of tacit knowledge.

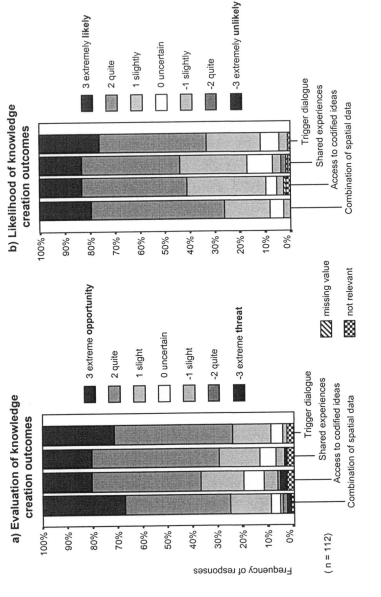

Figure 6.9   Assessment of knowledge creation outcomes

Perceived outcomes for inter-organisational relationships The last group of strategic position beliefs assessed the outcomes of sharing in terms of an organisation's relationship with other organisations. If sharing were to increase the interdependence among organisations or redistribute influence among organisations, 40-44% of respondents considered this an opportunity and less than 10% saw it as a threat to their strategic position (see Figure 6.10a).

Similarly, more than 40% of respondents considered these outcomes very likely (see Figure 6.10b). In sum, the outcomes of spatial data sharing for their organisation's strategic position were perceived in dissimilar ways. In terms of the information shared in the form of spatial data sets, the proportion of respondents perceiving various aspects of losing control over spatial data as a threat was generally larger than those perceiving them as opportunities. Yet the consequences of sharing in terms of the creation and transfer of tacit and explicit knowledge were evaluated very positively for their organisations' strategic position by the majority of respondents. Interestingly, the creation of explicit knowledge as an outcome of their organisations' sharing activities was valued by more respondents in the sample than the creation of tacit knowledge. Finally, more respondents perceived outcomes of sharing in terms of inter-organisational relations as an opportunity rather than a threat to their organisations' strategic position.

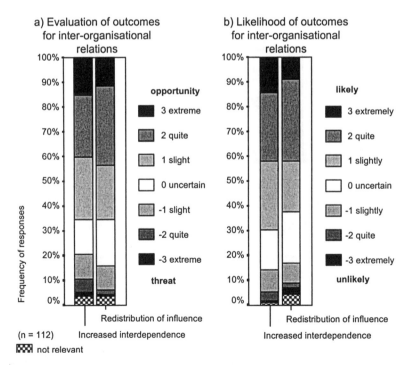

Figure 6.10 Assessment of outcomes for inter-organisational relations

Perceived Social Outcomes

The last domain under 'attitude' is social outcomes. Since the beliefs in this domain - referring to benefits for society at large and integrated development planning - are inherently positive, the positive or negative evaluation of these beliefs was not deemed necessary. Respondents were asked only to indicate the likelihood that these benefits might occur. Integrated development planning benefits from sharing spatial data were perceived as highly likely by more than 32% of respondents (see Figure 6.11). About 40% of respondents indicated that greater benefits to society at large were highly likely to result from sharing.

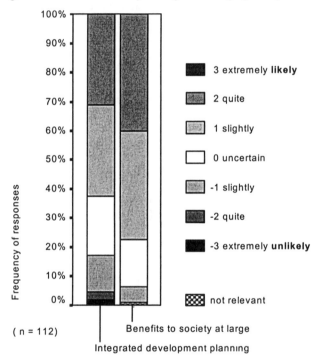

Figure 6.11 Perceived likelihood of social outcomes

In sum, specific social outcomes of spatial data sharing across organisational boundaries were considered very likely by only about one third of the respondents.

Social Pressure - Perceived Pressures to Share

'Social pressure' captures the expectations or pressures from important referents to engage or not to engage in spatial data sharing, as perceived by the organisation.

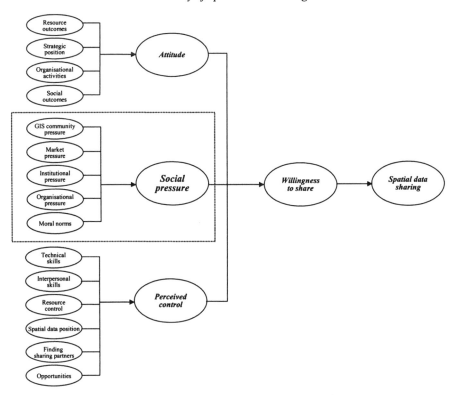

Figure 6.12 Model of the willingness to share spatial data: focus on social pressure and underlying domains

The conceptual location of the 'social pressure' component and its underlying domains is presented in Figure 6.12. The means for the direct measure for 'social pressure' and the domain level variables are presented in Table 6.3.

The mean score for the current 'social pressure' variable indicates that the general 'social pressures' to share spatial data across organisational boundaries were perceived to be only fairly strong. The mean score for the near future measure of 'social pressure' shows that pressures to share were expected to increase slightly to 'quite strong'.

At the domain level, the mean scores for the direct measures denote that the pressures to engage in sharing from the organisation and moral norms were perceived to be only marginally strong, whereas the market and institutional pressures were perceived to be a little weak while pressures from the GIS community were perceived to be neither particularly weak nor strong. The mean scores for the 'near future' measures at the domain level indicate that respondents expected an increase in the pressures to engage in spatial data sharing from the different groups of referents.

Table 6.3   Direct measures for social pressure and underlying domains

Direct measures	Current		Near future	
	Mean	**Standard deviation**	**Mean**	**Standard deviation**
Social pressure	4.72	1.33	5.49	1.29
Domain-level				
GIS community pressure	3.99	1.54	4.89	1.57
Market pressure	3.59	1.54	4.48	1.68
Institutional pressure	3.63	1.45	4.58	1.61
Organisational pressure	4.45	1.41	5.37	1.31
Moral pressure	4.76	1.46	5.67	1.33

Note: Responses have been rescaled to '1- extremely weak' to '7 - extremely strong' level
of pressure.
(n = 112)

The results of assessing the normative beliefs underlying 'social pressure' are presented for each of the five domains in the following sub-sections. Normative beliefs were assessed with respect to the *likely approval or disapproval* of spatial data sharing by the different groups of referents and the *general motivation to comply* with the expectations of each referent.

Perceived GIS Community Pressure

The first of the 'social pressure' domains is the GIS community. Figure 6.13 depicts how respondents perceived the expectations from various sectors in the GIS community regarding their organisation's engagement in spatial data sharing. For each sector, the respondents indicated whether their own organisation's involvement in sharing would be looked upon positively or negatively by the referents in the sector.

Generally, all of the sectors were perceived to be in favour of sharing, with 50% of the respondents or more indicating highly positive expectations from the sectors. Between 60-70% indicated that they perceived the different levels of government and academic research institutions as approving of their organisation's sharing activities. Somewhat fewer, 50-58% of respondents indicated that they deemed GIS users in the private sector, para-statal organisations, and non-governmental organisations to approve of spatial data sharing. However, a considerable share of respondents (16%) also indicated that the questionnaire item regarding the expectations of NGOs was not relevant.[37]

---

[37] The large number of respondents indicating the question regarding NGOs was not relevant suggests that NGOs were not an important referent in the GIS community. This corresponds with the absence of many GIS users in NGOs in South Africa at the time of the survey.

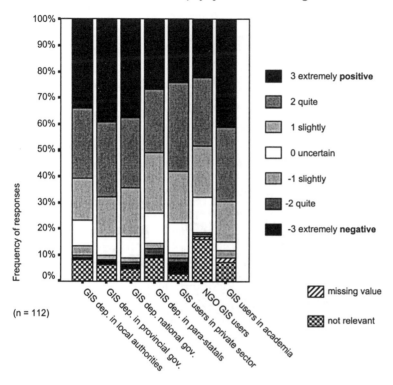

Figure 6.13 Perceived GIS community expectations

A pattern similar to the perceived expectations of the GIS community referents is shown with regard to the willingness of respondents' organisations to comply generally with these referents' expectations (see Figure 6.14). A high degree of motivation to comply with GIS departments in local, provincial and national governments and academic research institutions was indicated by 36-48% of respondents, whereas only 24-27% were motivated to comply with GIS departments in para-statal organisations, the private sector and NGOs. A considerable share (between 7 and 14%) of respondents considered this questionnaire item as not relevant.

In summary, all of the sectors in the GIS community were perceived to be in favour of sharing but the percentage of respondents indicating that their organisations were highly motivated to comply with the expectations of these referents was not very large.

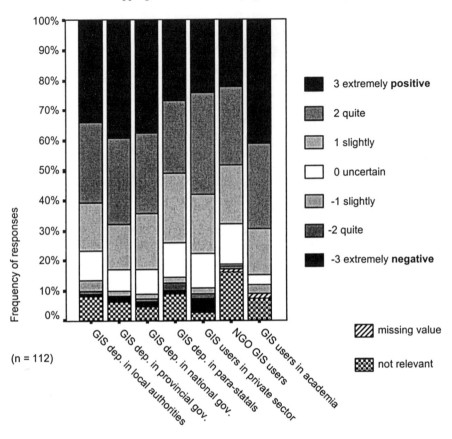

Figure 6.14 Motivation to comply with GIS community expectations

Perceived Market Pressure

The second domain under 'social pressure' consists of market pressures. Public spatial data brokers and GIS suppliers were perceived by more respondents (53% and 47% respectively) to approve of an organisation's engagement in sharing than were commercial spatial data brokers and private spatial data providers (36% and 38%, respectively) (see Figure 6.15a).

The motivation to comply with the different GIS market players was indicated to be low (see Figure 6.15b). While 30% of the respondents generally were motivated to comply with the expectations of public spatial data providers, only 10-14% indicated they would do so with commercial data brokers, private spatial data providers, and GIS suppliers. In contrast, up to 25% indicated they would definitely *not* be motivated to comply with these market players. A considerable number of respondents (10-15%) deemed these questions not to be relevant.

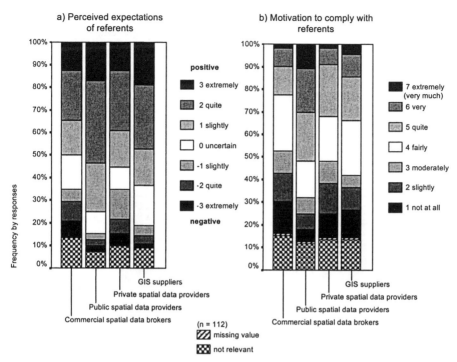

Figure 6.15 Perceived market expectations and compliance

In summary, the scores on the beliefs about market pressure provide signs that these various stakeholders in the spatial data market do not constitute important referents for the decisions of individuals within organisations to engage in spatial data sharing.

Perceived Institutional Pressure

The third 'social pressure' domain is institutional pressure. Figure 6.16a shows the respondents' perception of how their organisation's engagement in spatial data sharing is regarded by certain institutions. About 60% of respondents perceived that their engagement in sharing would be regarded very positively by the NSIF. Slightly less than half indicated that sharing was approved within the framework of existing spatial data agreements of their organisation. Finally, just over 40% expected that politicians would approve of their organisation's sharing activities. For all three institutional beliefs, a relatively large number of respondents (17-22%) regarded the question as being not relevant.

Figure 6.16b presents the motivation to comply generally with the expectations of these institutions. It shows that around half of the respondents were generally motivated to comply with the NSIF and with spatial data agreements of their organisation. Only 35% of the respondents indicated they were very motivated to comply with politicians. Again, for all three beliefs, a relatively large percentage of respondents (10-21%) indicated that the question was not relevant.

Overall, among the institutional pressures that were examined, the NSIF emerged as the most important referent for the decision of organisations to engage in spatial data sharing activities. However, the large number of respondents indicating the NSIF-related questions to be 'not relevant' (10-17%) possibly is a reflection of the lack of awareness of the NSIF and its activities within the GIS community.

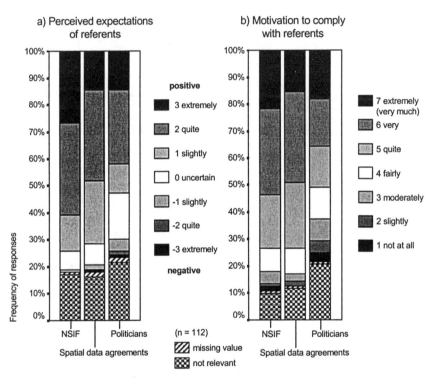

Figure 6.16 Perceived institutional expectations and compliance

Perceived Organisational Pressure

The fourth domain within 'social pressure' is organisational pressure. The organisational pressures regarding an organisation's engagement in sharing spatial data were perceived to be towards sharing by the majority of respondents (see Figure 6.17a). Other departments within the respondents' organisation were expected to regard their organisation's sharing activities as very positive by 66% of respondents. Similarly, 64% expected extremely positive reactions to sharing spatial data from their management and 71% from 'champions' for sharing in their organisation. The motivation to comply with the three referent groups was high for 49-61% of the respondents (see Figure 6.17b).

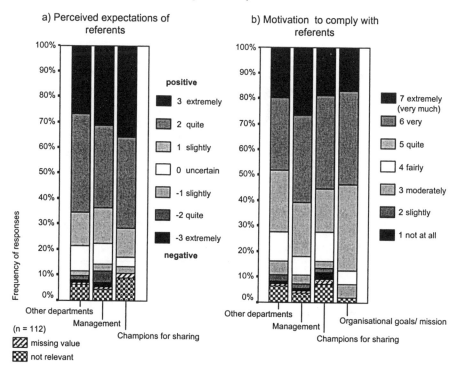

Figure 6.17 Perceived organisational expectations and compliance

Figure 6.18 summarises how respondents perceived their organisation's mission or goal with regard to their organisation's engagement in spatial data sharing. A large majority (87.5%) indicated that their organisational goals suggested they should share, only 6.2% indicated their mission suggested they should not share spatial data across organisational boundaries, while 5.4% of respondents gave no indication either way and marked the question as not relevant.

As shown in Figure 6.17b, almost 55% of respondents perceived the general compliance of people in their organisation to be very high with the organisational goals and mission.

In sum, the referents within their own organisation, including their organisational goals or mission, were perceived to be in favour of spatial data sharing by the majority of the respondents. The indicated motivation to comply with the expectations of these referents was high for more than half of the respondents.

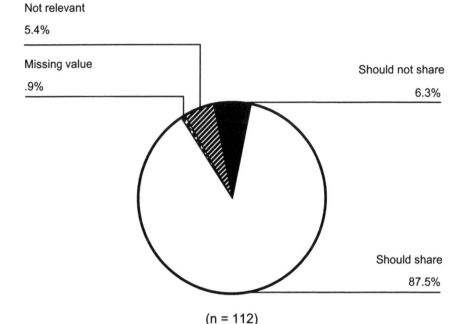

Not relevant

5.4%

Missing value

.9%

Should not share

6.3%

Should share

87.5%

(n = 112)

**Figure 6.18 Perceived organisational goals / mission regarding spatial data
             sharing**

Perceived Moral Norms

The last domain under 'social pressure' is moral norms. The moral obligation for
an organisation to engage in sharing spatial data was assessed in relation to two
separate social beneficiaries. For 62% of respondents, integrated development
planning was perceived to imply that their organisation's engagement in sharing
would be highly desirable (see Figure 6.19a). About half the respondents perceived
that society at large would favour their organisation's sharing activities. However,
more than 10% considered the questions to be not relevant. General compliance
with these norms was measured in terms of the willingness of the respondents'
organisation to engage in spatial data sharing just for the sake of these social
beneficiaries and around 50% of the respondents reported to be very willing
(Figure 6.19b).

    In summary, the results indicate that for more than half of the respondents, the
social beneficiaries are important referents for their organisations' decision to
engage in spatial data sharing activities.

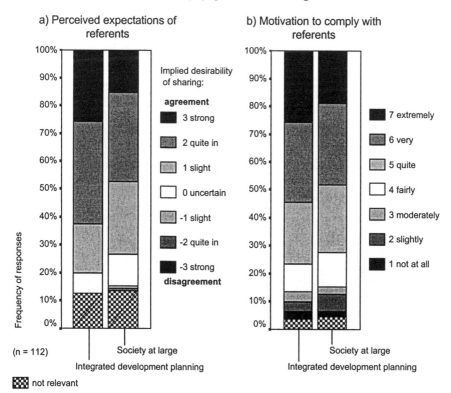

a) Perceived expectations of referents

b) Motivation to comply with referents

Frequency of responses

(n = 112)

Implied desirability of sharing:

agreement

■ 3 strong

▨ 2 quite in

□ 1 slight

□ 0 uncertain

▨ -1 slight

▨ -2 quite in

■ -3 strong

disagreement

■ 7 extremely

▨ 6 very

□ 5 quite

□ 4 fairly

▨ 3 moderately

▨ 2 slightly

■ 1 not at all

Society at large

Integrated development planning

▨ not relevant

Figure 6.19 Perceived moral norms and compliance

Perceived Control over Sharing

'Perceived control' measures how easy or difficult the performance of spatial data sharing is for an organisation taking account of the presence or absence of requisite resources and skills. The conceptual location of the 'perceived control' component and its underlying domains within the model of the willingness to engage in spatial data sharing is highlighted in Figure 6.20.

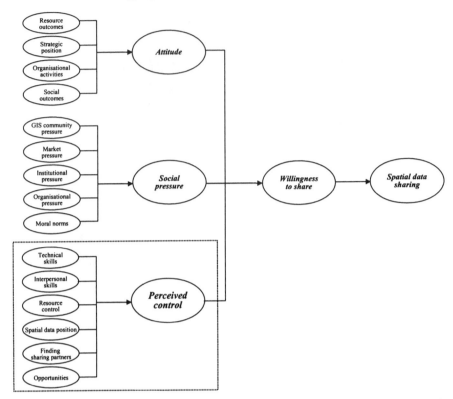

Figure 6.20 Model of the willingness to share spatial data: focus on perceived control and underlying domains

The mean scores of the direct measures for 'perceived control' and the domain level variables are presented in Table 6.4. The mean scores for the *current* variable of 'perceived control' indicate that the organisations of the respondents in the sample were perceived to be in control over sharing only to a small extent. The mean scores for the *near future* measure of 'perceived control' show that respondents expected the degree of control over sharing to increase slightly.

A comparison of the mean scores at the domain level indicates that the current control over sharing in these domains was perceived to be somewhat lower than in the near future. The current control over technical and interpersonal skills was perceived to be quite high, while control in terms of resources and the spatial data position of the organisations was perceived to be only moderate. The mean scores for the opportunities domain indicates that respondents perceived their organisations to be only fairly reliant on opportunities for sharing spatial data with other organisations and that they expected a decrease in reliance in the near future.

Table 6.4   Direct measures for perceived control and underlying domains

Direct measures	Current		Near future	
	Mean	Standard deviation	Mean	Standard deviation
Perceived control	4.91	0.96	5.34	0.91
Domain-level				
Technical skills	5.13	1.70	5.60	1.58
Interpersonal skills	5.11	1.49	5.71	1.10
Resource control	3.77	1.68	4.69	1.63
Spatial data position	3.59	1.47	4.29	1.50
Finding sharing partners *				
Opportunities	3.66	1.69	2.53	1.69

Note: Responses have been rescaled to '1 - extremely low' to '7 - extremely high' level of control.
* 'Finding sharing partners' was not addressed with a direct measure at the domain level in the questionnaire.
(n = 112)

'Perceived control' is assumed to be based on a number of control beliefs. Control beliefs are concerned with the requisite resource, skills and opportunities for spatial data sharing. They were assessed firstly, with regard to the perceived *importance* of each skill or resource for carrying out spatial data sharing activities across organisational boundaries and secondly, with respect to the perceived extent of *control* over each resource or the perceived availability of each skill. The results of assessing the six domains of control beliefs are presented below.

Perceived Control over Technical Skills

The technical skills domain of 'perceived control' taps beliefs in three areas; spatial data-related skills, metadata-related skills and more general information technology skills (for a justification of each area, see Chapter 5).

Spatial data skills   The spatial data-related skills were deemed important for sharing activities by many respondents (see Figure 6.21a). Those that were considered important by most respondents were assessing the quality of spatial data (indicated by 86%) and integrating spatial data from diverse sources (indicated by 84%). These skills received particular emphasis as none of the respondents indicated that these skills were not important. Handling different formats of spatial data was considered important by 70% of the respondents, mastering different standards by 53% and selecting spatial data from a database by 52% of respondents.

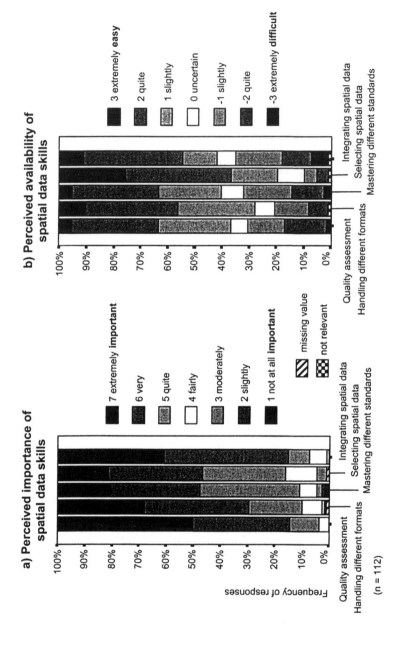

Figure 6.21 Spatial data skills

Considering the importance attributed to the spatial data-related skills, the number of respondents reporting a high skills level in this area in their organisation was quite small (Figure 6.21b). Only 37-47% of respondents reported that quality assessment, handling different formats, mastering different standards and the integration of spatial data from different sources was easy for their organisation (versus 7-17% who regarded it as difficult). The exception to this was selecting spatial data from a database which was considered easy by 60% of the respondents.

Metadata skills   Similarly, although metadata-related skills were regarded important by many respondents (see Figure 6.22a), a much smaller percentage of respondents indicated to possession of these skills (see Figure 6.22b).

Maintaining and updating metadata were considered important by 71% of the respondents and 32% indicated they were easy for their organisation. Capturing metadata was regarded an important skill for sharing by 64% but only 37% indicated they regarded capturing metadata easy for their organisation (versus 11%). Applying metadata standards was perceived important by 56% (versus 8%) while this skill was easy for only 30%. Rated at the same level of importance was the ability to interpret metadata which almost 40% of the respondents indicated their organisation could do easily (versus 6%). Particularly low levels were indicated for using metadata interfaces and catalogues; only 30% (versus 10%) reported this was easy for their organisation, while 45% indicated that the ability to use these interfaces was important for sharing spatial data.

IT skills   The last area of technical skills assessed is general information technology skills. Among these skills, database administration and the transfer of spatial data to and from different media were considered particularly important by 64% and 70% of the respondents respectively (see Figure 6.23a). Only 41% considered using the Internet to locate spatial data and 53% to distribute spatial data to be important skills for sharing spatial data.

In terms of the ease or difficulty of carrying out these activities, the transfer of spatial data to and from different media was considered easy by 77% of the respondents (see Figure 6.23b). Yet only 44% considered database administration easy. Using the Internet to *locate* spatial data was easy for the organisations of 57% of the respondents while using the Internet to *distribute* spatial data was easy for only 45%.

In summary, these results indicate that for all three areas of technical skills (spatial data, metadata and IT), the perceived skills level in the organisations does not match the importance attributed to these skills for sharing spatial data.

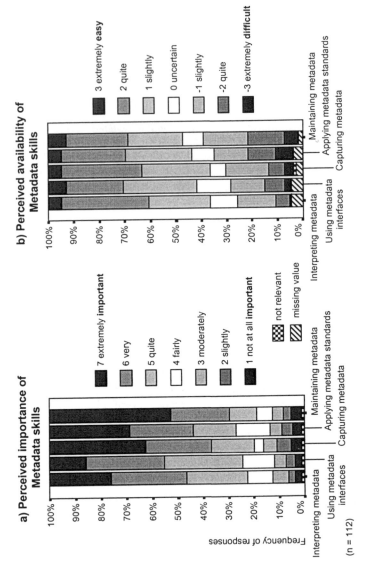

Figure 6.22 Metadata skills

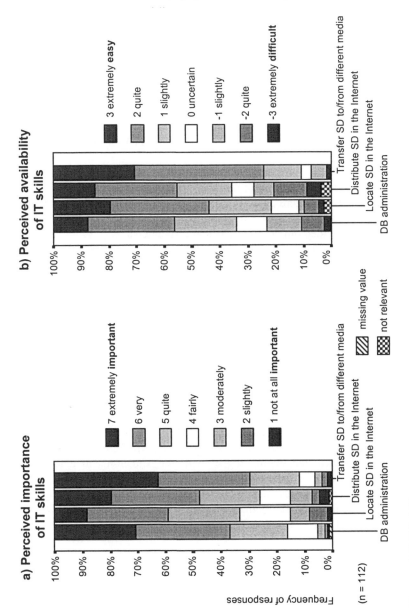

Figure 6.23 IT skills

Perceived Control over Interpersonal Skills

The second domain under 'perceived control' is interpersonal skills. The interpersonal skills domain was assessed by considering control beliefs from three different areas; networking and teamworking skills, negotiation skills, and past experience.

Networking and teamworking skills Networking and teamworking skills were considered important for spatial data sharing by the majority of respondents (see Figure 6.24a).
Networking skills, consisting of establishing and fostering a network of contacts, identifying and attending meeting opportunities and keeping a finger on the pulse of a network, were perceived to be very important for sharing by 61-74% of the respondents. Similarly, collaborating with others and interdisciplinary teamwork were deemed important by 70% of respondents.
Regarding the perceived ease or difficulty of these activities, just over 40% of respondents indicated that networking and collaboration were very easy activities for their organisations, with the exception of keeping a finger on the pulse of a network (see Figure 6.24b). Less than a quarter of respondents regarded this as easy and 12% of the respondents even considered it difficult for their organisation.

Negotiation skills The second area of interpersonal skills is negotiation skills (see Figure 6.25a). Negotiating the different characteristics of any spatial data sharing agreement such as a 'win-win' situation, spatial data pricing, and liability, were deemed important for sharing spatial data by more than 60% of the respondents. Being able to negotiate data ownership agreements such as copyright was very important for 73% of respondents.
Again, the self-reported skills levels were ranked much lower than the perceived importance of these skills (see Figure 6.25b). Only around 40% of the respondents reported that negotiating a 'win-win' situation and data ownership agreements was easy for their organisation, compared to 12-15% who considered this difficult. Even fewer respondents (25-30%) indicated that negotiating the price of spatial data and liability agreements was easy for their organisation while about 20% reported that this was very difficult for their organisation.

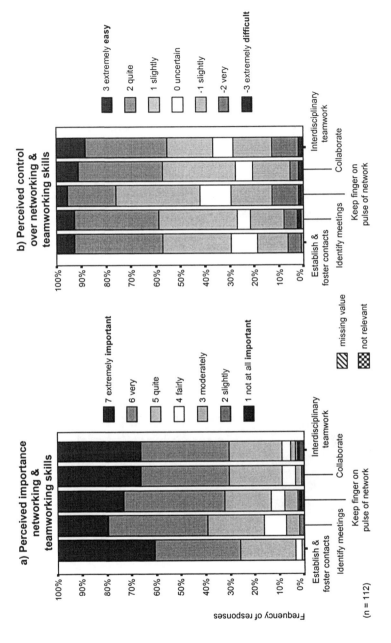

Figure 6.24 Networking and teamworking skills for spatial data sharing

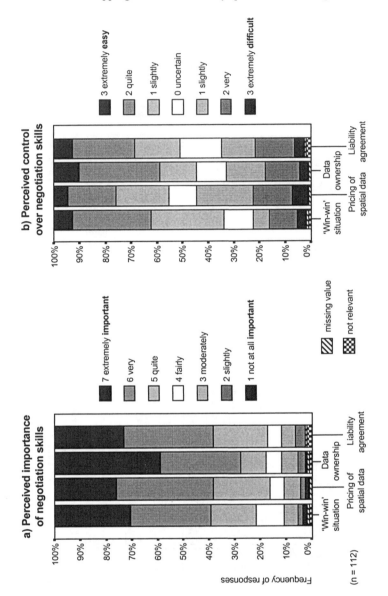

Figure 6.25 Negotiation skills for spatial data sharing

Past experience   The reports of their organisation's past experience with spatial data sharing were quite mixed (see Figure 6.26). Extensive experience with sharing spatial data with other organisation in the past was indicated by 42% of respondents while just over 20% indicated they could not build on such experience. Furthermore, about one fifth of respondents reported that their organisations had bad experiences with sharing across organisations. Finally, just over one third of respondents claimed to have a clear overview of the consequences of sharing spatial data for their organisation while 22% indicated they were lacking a clear overview.

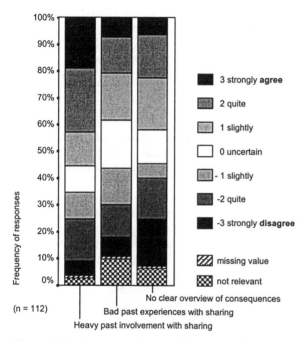

Figure 6.26 Past experience with spatial data sharing

Overall, while interpersonal skills in networking, teamworking, and negotiation were deemed important by the majority of the respondents for spatial data sharing activities across organisational boundaries, the perceived skills level in their organisations was considered to be limited. Similarly, past experience with spatial data sharing among the organisations represented by the respondents was limited.

Perceived Control over Resources

The third domain under 'perceived control' is resource control. The beliefs in this domain pertain to the resources that may be necessary in order to engage in spatial data sharing. Of central concern is the extent to which those resources deemed essential are available and under control of the organisation.

Between 53% and 65% of the respondents indicated that a sufficient number of staff, time to spend on sharing activities, and the availability of funding are highly important for sharing, but 70% of the respondents considered organisational guidelines related to sharing to be important (see Figure 6.27a).

While the resources, particularly funding and organisational guidelines, were deemed necessary for sharing, the availability of these resources was perceived to be very low (see Figure 6.27b). A lack of control over funding and organisational guidelines related to spatial data was reported by nearly 40% and 30% of the respondents, respectively, while these resources were perceived to be available to, and under control of, less than 20% of the respondents. Almost 45% of the respondents reported very low control over staff numbers compared to about 5% with high control over having sufficient staff for sharing activities. Similarly, more than 30% indicated their organisation did not have time available to spend on sharing activities while less than 10% of the respondents indicated that their organisation did have control over allocating sufficient time to sharing activities.

Spatial Data Position

The fourth domain under 'perceived control' is the spatial data position. The position of an organisation with respect to spatial data entails the extent of both the control over internal spatial data resources and the dependence on external spatial data sources (see also Chapter 5).

Perceived control over internal spatial data One aspect of the spatial data position of an organisation is the control over its own spatial data. More than 40% of respondents perceived their organisation's spatial data as very important for their sharing partners (see Figure 6.28a). Between 30-40% of respondents perceived that copyright, regulated access, controlled use, and the enforcement of rules are very important aspects of controlling their own spatial data resources.

The number of respondents who perceived that their organisation was able to implement these control aspects was quite small (only 20-25% of respondents) (see Figure 6.28b). Especially low (18%) was the number of respondents who believed they *could* ensure that spatial data owned by their organisation could *not* be used by organisations other than their sharing partners (compared to about 35% who believed their organisation could not ensure this). A third of respondents indicated that the extent to which their organisation was able to enforce rules regarding copyright of, access to, and use of, its spatial data was extremely small.

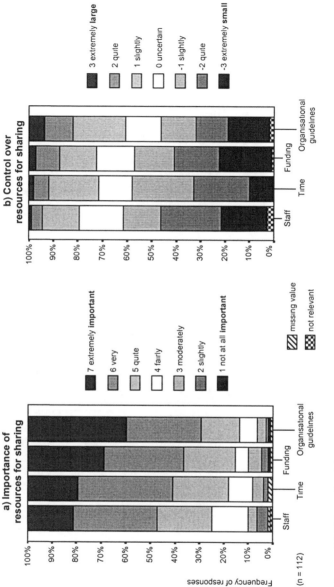

Figure 6.27 Perceived importance of, and control over, resources for spatial data sharing

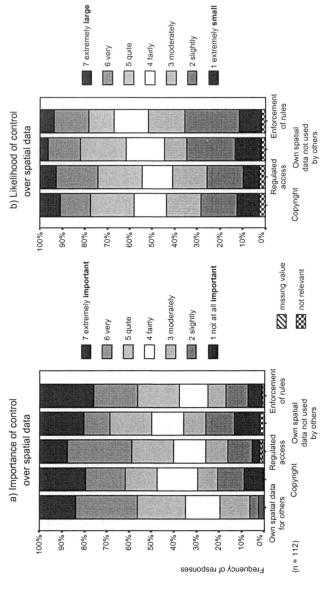

Figure 6.28 Perceived control over spatial data

Further aspects of the control over an organisation's internal spatial data resources are the availability and stability of alternative sources (see Figure 6.29b). Less than 20% of respondents believed that there were many alternative sources for their organisation's spatial data while more than 35% indicated that the number of alternatives was very limited. Equal numbers of respondents (about 20%) considered the alternative sources for their organisations' spatial data to be either very stable or very unstable. By almost 10% of respondents considered these questions to be not relevant.

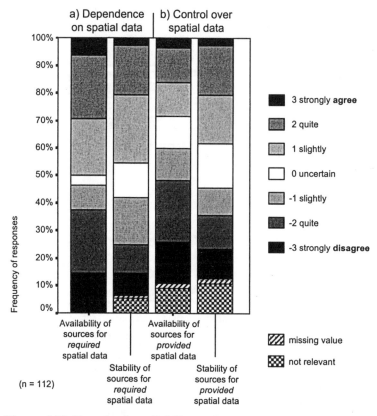

Figure 6.29   Perceived availability and stability of alternative sources of spatial data

Perceived dependence on external spatial data   The second aspect of an organisation's spatial data position is its dependence on external sources of spatial data. Half of the respondents believed that the spatial data their organisation would receive from external sources would be very important for the continued operation of their own organisation. This compared to 12% who believed their organisation's operation would not be affected (see Figure 6.30a).

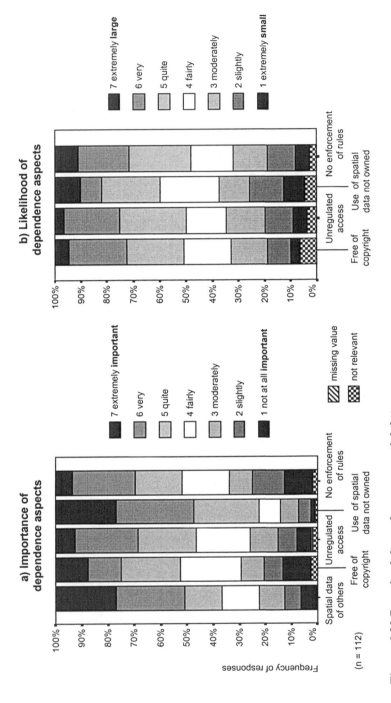

Figure 6.30 Perceived dependence on spatial data

Dependence was measured by the extent of control of other organisations. Among the aspects of spatial data control by other organisations, 53% of respondents indicated that the ability to use spatial data not owned by their organisation was highly important. Between 25-30% indicated that it was highly important to their organisation that received spatial data was free of copyright, that access to required spatial data was unregulated, and there was no enforcement of rules regarding the copyright of, access to, and use of external spatial data.

The number of respondents expecting that other organisations would be able to enforce these control aspects for their spatial data was reasonably small (see Figure 6.30b). Just under 30% of the respondents perceived that other organisations would be able to effectively enforce copyright, regulate access to their spatial data, and enforce rules regarding their spatial data. Even fewer (18%) believed that other organisations would be able to ensure that the respondents' organisation could not use spatial data owned by other organisations.

Regarding the availability of external sources of spatial data, 37% of respondents indicated that their organisations do not have many alternative sources for the spatial data they require while 30% believed they did have alternative supplies (see Figure 6.29a). Concerning the stability of the alternative sources that do exist, equal shares of respondents (about 20%) believed these sources to be either very stable or very unstable.

In terms of the dependence on other organisations for spatial data, a quarter of the respondents perceived their organisations to be very dependent while another quarter considered their organisation to be highly self-sufficient (see Figure 6.31a).

With respect to *giving* versus *receiving* spatial data, about 45% of the respondents reported that their organisations' position was more or less balanced, 24% of respondents reported to be mainly *giving* spatial data (to a greater or lesser extent), and just under 32% of the respondents reported that their organisation was *receiving* data (to a greater or lesser extent) (see Figure 6.31b).

In summary, in terms of control over internal spatial data, the share of respondents indicating that their organisation was able to implement the means of control was small. Similarly, in terms of their own organisations' dependence on external spatial data, other organisations were not perceived to be able to enforce the control aspects and only one quarter of the respondents indicated their organisation to be very dependent on other organisations for spatial data.

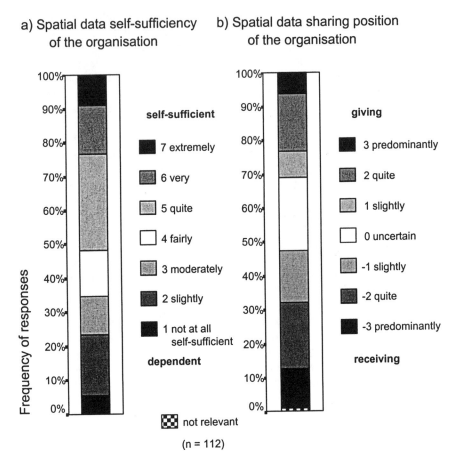

Figure 6.31　Perceived spatial data sharing position

Perceived Control over Finding Sharing Partners

The fifth domain under 'perceived control' is finding sharing partners. The previous domains under 'perceived control' were considering the necessary skills and resources for spatial data sharing across organisational boundaries that are internal to the organisation. This domain captures one of the external factors upon which organisations may depend to engage in spatial data sharing: the dependence on the co-operation of other people or organisations to engage in spatial data sharing. From Figure 6.32a, it is evident that the willingness and responsiveness, and the reliability of sharing partners were considered very important characteristics by 78% and 85% of the respondents, respectively. In contrast, a compatible purpose of application and organisational fit were important characteristics for current or prospective sharing partners for only 47% and 34% of the respondents. Only a small number of respondents expected that finding sharing partners with these characteristics would be very easy (see Figure 6.32b), while just over 30% of the respondents expected to easily find willing, responsive, and reliable sharing partners. Twenty four % regarded finding sharing partners with a compatible purpose of application and 20% with a compatible organisational fit to be easy for their organisation (compared to 10% who considered it difficult).

In sum, the most important characteristics of potential sharing partners were deemed to be willingness and responsiveness as well as reliability. Yet only a small percentage of the respondents perceived that finding sharing partners with any of these characteristics was very easy for their organisations.

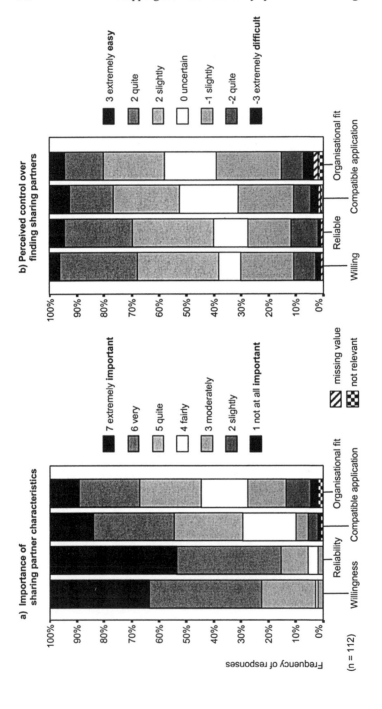

Figure 6.32 Perceived control over finding sharing partners

Perceived Opportunities for Sharing Spatial Data

The last domain under 'perceived control' is the opportunities domain. The role of facilitating and providing opportunities for spatial data sharing of the National Spatial Information Framework (NSIF) was assessed by relating the core NSIF activities to the spatial data sharing activities of the respondents' organisation.

More than 70% of respondents perceived the identification of core data sets and the spatial data discovery facility as very important for their organisation's sharing activities (see Figure 6.33). For 58% of respondents, standards alignment was very important and slightly fewer respondents considered policy development activities (48%) and NSIF awareness creation (43%) very important for sharing. The fora organised by the NSIF were important to only 36% of respondents and, for a further 10%, the question was not relevant.

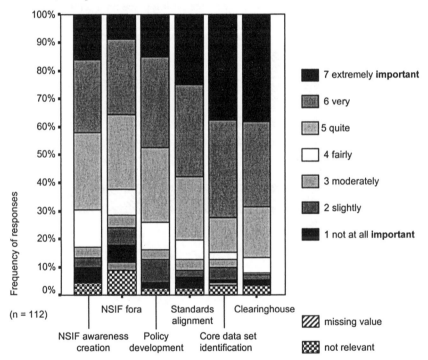

Figure 6.33 Perceived opportunity creation by the NSIF

Overall, the results for the items used to assess the scope and relevance of the NSIF activities as creating opportunities for spatial data sharing suggest that the identification of core data sets, the provision of the spatial data facility and the alignment of standards were perceived to be most important.

Self-reports of Spatial Data Sharing

Self-reports of actual spatial data sharing were included in the questionnaire to respond to the second research question regarding the extent of actual spatial data sharing behaviour that is already taking place (see Chapter 5 for an explanation of the measures used to assess spatial data sharing activities).

The magnitude of actual spatial data exchanges among the sample respondents is indicated by the frequency, schedule and recency of reported spatial data exchanges, and by the average quantity of spatial data that is reported to have been shared.

As shown in Figure 6.34, only a small minority of respondents reported that their organisation had never engaged in spatial data sharing activities at all. The remaining responses are not mutually exclusive. Respondents indicated that by far the most frequent spatial data exchanges were taking place on a project basis.

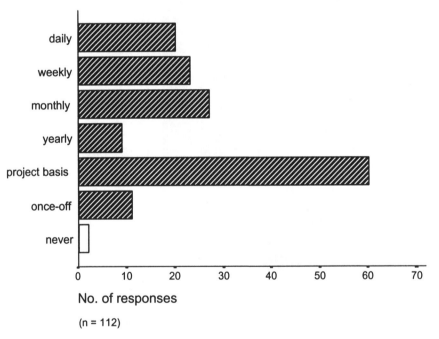

(n = 112)

Figure 6.34 Frequency of spatial data exchanges

Figure 6.35 distinguishes the frequency of spatial data exchanges according to the different GIS community sectors. It is important to note that respondents were able to indicate multiple frequencies (unless they indicated 'never' to sharing); hence the total number of responses per sector may exceed the total number of respondents per sector. From Figure 6.35, it is possible to detect the sectors in the sample that indicated their organisation was not engaging in any sharing activities. These respondents were from the private sector and from para-statal organisations.

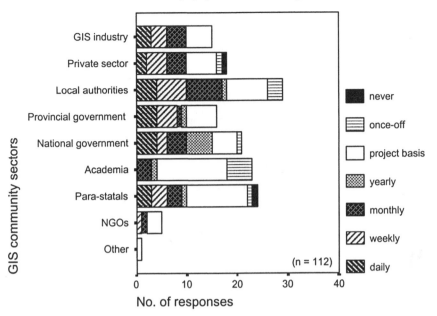

Figure 6.35 Frequency of spatial data exchanges by sector

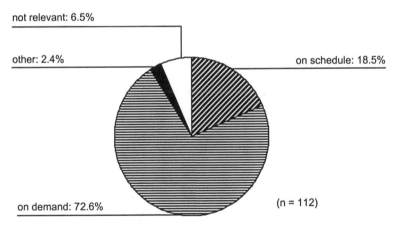

Figure 6.36 Schedule of spatial data exchanges

The schedule of spatial data exchanges reported by the respondents is indicated in Figure 6.36. It shows that the majority of exchanges across organisational boundaries were reported to have been on demand (72.6%) and that less than a fifth took place on a scheduled basis (18.5%).

Figure 6.37 presents the reported schedule of spatial data exchanges by sector. It is noticeable that more respondents from local authorities and para-statal organisations than from any of the other sectors reported that exchanges took place on a scheduled basis.

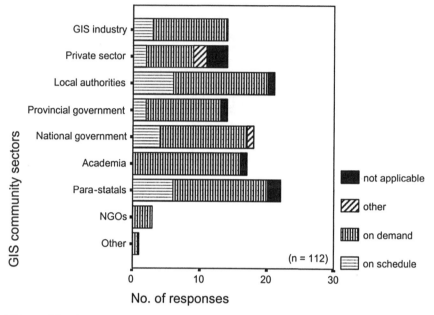

Figure 6.37 Schedule of spatial data exchanges by sector

Respondents were also asked to indicate when the most recent spatial data exchanges of their organisations had taken place. Figure 6.38 shows that more than three-quarters of the exchanges had taken place within the last month or more recently.

Details of the average quantity of spatial data exchanged by organisations as indicated by the respondents in the sample are presented in Figure 6.39. Almost half of the respondents reported that entire spatial data bases were exchanged. Frequent options were also to exchange a selected subset or a thematic layer of spatial data. There were far fewer indications of the exchange of single or summary observations.[38] For additional details and analysis of the self-reported sharing behaviour in the sample, see Appendix D.

---

[38] Summarising spatial data (such as classifying various types of vegetation under one title) limits the amount of information disclosed.

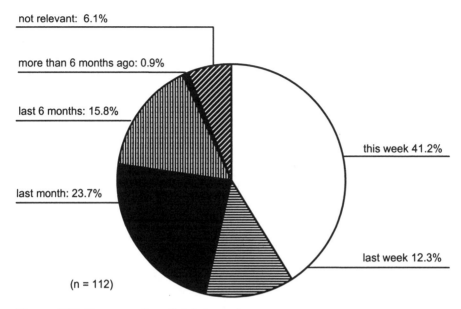

not relevant: 6.1%

more than 6 months ago: 0.9%

last 6 months: 15.8%

last month: 23.7%

(n = 112)

this week 41.2%

last week 12.3%

Figure 6.38 Recency of spatial data exchanges

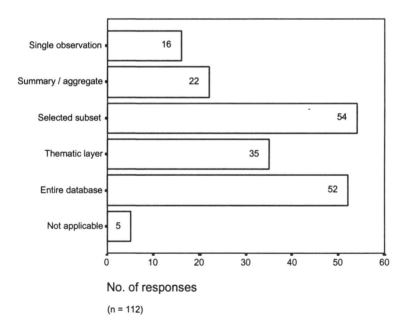

No. of responses

(n = 112)

Figure 6.39 Average quantity of spatial data exchanges

As discussed in Chapter 3, the scope of spatial data sharing among organisations can be categorised according to the following levels (Obermeyer and Pinto, 1994; Pinto and Onsrud, 1995): a) situation specific, project driven, non-routine, and non-recurring, to solve specific problems; b) regular sharing and exchange of information using protocols and procedures for limited sharing parties; and c) routine sharing, standardised, generalisable patterns of exchange, accessible to all parties in terms of location and format.

The magnitude of spatial data sharing reported by the respondents in the sample is characterised by exchanges on demand rather than scheduled exchanges, taking place mainly on a project basis. The exchanges were reported to have taken place recently and, according to many respondents, the average quantity of spatial data exchanged by their organisation entailed the entire database, a subset or a thematic layer. Despite the recency of reported exchanges, the self-reports of spatial data sharing of the respondents in this sample indicate that only the first level of ad hoc, non-routine level of sharing had been reached at the time of the survey.

Conclusion

This chapter has presented the descriptive results of the survey. It has also addressed the first research question regarding the willingness of organisations to engage in spatial data sharing. The analysis of the survey data indicates that, overall, the willingness of the organisations in the sample to share spatial data across organisational boundaries was low, with a slight increase expected in the near future. It further indicates that the willingness of organisations, as expressed by the respondents in the sample, to engage in spatial data sharing across organisational boundaries did not vary for the different sectors in the GIS community, nor did it differ with the extent of perceived self-sufficiency or dependence of organisations on spatial data or the organisation's spatial data position in terms of predominantly giving spatial data to, or receiving spatial data from, other organisations.

Addressing the second research question, self-reports of actual sharing activities revealed that, although most respondents indicated some involvement of their organisations in spatial data sharing across organisational boundaries, the degree of spatial data sharing activities was still very limited.

The analysis required to validate the results of the questionnaire and the underlying model is summarised in Chapter 7, together with an analysis of the most salient determinants of the perceived willingness to engage in spatial data sharing across organisational boundaries.

To Share or Not to Share: Determinants of the Willingness to Engage in Spatial Data Sharing

Introduction

The principal research question in this work is concerned with identifying the factors that influence key individuals within organisations to engage in spatial data sharing across organisational boundaries. In answering this question, the research is expected to identify the existing incentives and disincentives to their participation in spatial data sharing activities. Improved understanding of these incentives and disincentives is expected to provide a more effective basis for fostering spatial data sharing to overcome bottlenecks in the availability of, and access to, spatial data.

The purpose of this chapter is to present the analysis of the questionnaire-based survey data collected in Phase II of the empirical research. In the first section, the results of the model validation against empirical data is discussed. Next, the specific determinants of the willingness of organisations in South Africa to share spatial data are identified. How these determinants can be addressed in order to foster a spatial data sharing 'culture' is considered in the next section. The chapter concludes with a summary.

Empirical Validation of the Model

> When a measure is devised, it should not be presumed to be appropriate and adequate; rather, it is necessary to establish that it meets the researcher's aims and that it has adequate measurement properties. (Bryman, 1989: 54)

In order to conduct the analysis, the model of the willingness of organisations to engage in spatial data sharing is validated using the empirical data (for details of the validation procedures and results, see Appendix E). The verification of the model shows that, in terms of the reliability of the measures, the items in the questionnaire had highly satisfactory levels of internal reliability. With respect to conceptual validity, the analysis of the empirical data confirms that the domains and the beliefs appear to have tapped the three concepts 'attitude', 'social pressure' and 'perceived control' as suggested by the model developed in Chapter 5. Finally, concerning construct validity, the relationships between the variables are all in the expected direction; with respect to the relationships between the sum of combined beliefs and the results for the direct measures of 'attitude', 'social pressure', and 'perceived control', and with respect to the explanatory power of the model.

The validation of the model demonstrates, firstly, that the finding in Chapter 6 regarding the low willingness of organisations to share spatial data is valid and that the model developed in this study can be used to address the research question regarding the determinants of the willingness of organisations to engage in spatial data sharing. Secondly, the model, as implemented using the questionnaire instrument, could be applied in other national contexts. The attainment of the validity of the model constitutes a crucial step in the approach of this study which proposes to systematise the determinants of an organisation's spatial data sharing behaviour. Only once a model has been demonstrated to be valid, can it be relied upon to further the understanding of the incentives and disincentives for key individuals within organisations to engage in spatial data sharing across organisational boundaries.

Analysis of the Determinants of the Willingness to Engage in Spatial Data Sharing

Having validated the model of the willingness to share spatial data, the next step in the analysis is to examine what the principal determinants of the perceived willingness to engage in spatial data sharing might be. As the model consists of several possible sources of willingness, the appropriate statistical method for establishing the most important ones is multiple regression analysis. Multiple regression provides a means of choosing empirically the most effective set of predictors of a variable (Howitt and Cramer, 1997) and of establishing the relative importance of each independent variable in the prediction (Bryman and Cramer, 1999).

Following the hierarchical structure of the model as presented Appendix E, the analysis proceeds from examining the importance of the main components ('attitude', 'social pressure', 'perceived control') to explain intention, to the identification of the determinants of each of the main components. Selected details of the regression outcomes are presented in the following sections and the complete regression outputs are presented in Appendix E.

Determinants of Intention

Analysis of the data set indicates that positive relationships exist between the three main components ('attitude', 'social pressure', 'perceived control') and the perceived willingness to share as indicated by the significant correlation coefficients presented in Table 7.1. For the case of 'perceived control', the coefficient is weaker but still significant. Hence, the more positively that the outcomes of data sharing are perceived to be, the more willing an organisation is likely to be to engage in spatial data sharing. The more 'social pressures' are perceived to be in favour of sharing, the higher the likely willingness of an organisation to share spatial data. Finally, the higher the 'perceived control' of an organisation over the requisite skills and resources for spatial data sharing, the higher the likely willingness to engage in spatial data sharing.

Table 7.1   Correlation between intention and main components

Construct	Pearson correlation coefficient (r) Direct measure of intention	
	current	near future
Attitude	.547**	.619**
Perceived social pressure	.640**	.663**
Perceived control	.204*	.354**

**. Correlation is significant at the 0.01 level (2-tailed)
*. Correlation is significant at the 0.05 level (2-tailed)
(n = 112)

However, in order to examine how 'attitude', 'social pressure' and 'perceived control' *jointly* influence the likely willingness to share, a multiple regression technique was applied (detailed results are presented in Appendix E). The multiple regression coefficient (R^2) indicates the collective effect of all the independent variables on the dependent variable (willingness to share) and serves as a measure of the extent to which willingness to share can be predicted from the three main components ('attitude', 'social pressure' and 'perceived control'). According to the results of the regression, 46% of the variation in the willingness to share can be explained ($R^2 = .46$; $F(3, 108)= 30.61$; $p<.000$) by the direct measures of 'attitude', 'social pressure' and 'perceived control'.

For each independent variable, the significance level of its coefficient indicates whether a variable makes a significant addition to the strength of the prediction of willingness to share over and above the contribution of the other two independent variables. Hence, the significance levels in Table 7.2 indicate that 'perceived social pressure' and 'attitude' ($p<.000$ and $p= .005$, respectively), but not 'perceived control' ($p= .265$), are significant predictors of willingness to share spatial data.[39]

---

[39] A high p value for the t test of an independent variable indicates that the probability of attaining the regression coefficient for that variable by chance is high, so that the regression coefficient should not be relied upon.

Table 7.2   Standardised regression coefficients with intention as dependent variable (*present*)

Independent variables	Standardised regression coefficient (Beta)	Significance p
Attitude	.255	**.005**
Perceived social pressure	.462	**.000**
Perceived control	.084	.265

Similarly, analysis of the measures for the near future[40] provides further support for the role of 'attitude' and 'social pressure' in explaining the willingness to share.[41] As indicated in Table 7.3, the significance levels of the coefficients for 'attitude' (p<.000) and 'social pressure' (p<.000) in the near future indicate that these components add significantly to the prediction of the willingness to share in the near future. And, again, the significance level of the coefficient for 'perceived control' (p= .574) in the near future means that 'perceived control' cannot be confirmed as a predictor of the willingness to share.

Table 7.3   Standardised regression coefficients with intention as dependent variable (*near future*)

Independent variables	Standardised regression coefficient (Beta)	Significance p
Attitude	.324	**.000**
Perceived social pressure	.434	**.000**
Perceived control	.043	.574

However, it has been argued that non-significant coefficients provide only weak evidence of the *absence* of an effect because the sample may not have been large enough to detect the effect (Allison, 1999; Bryman and Cramer, 1999). Indeed, as Table 7.6 shows, the simple relation between intention and 'perceived control' is significant (current: r= .204, p<.05; near future: r= .354, p<.01). Considered individually against intention, each of the three variables ('attitude', 'social pressure' and 'perceived control') is significant. However, when considered jointly, intention is more strongly associated with 'attitude' and 'social pressure' than with 'perceived control'. 'Social pressure' and 'attitude' can be said to have

---

[40] 'Near future' refers to the time period '*within the next month up to two years*' from the time of conducting the survey.

[41] The regression results show that 50% of the variation in the willingness to share in the near future can be explained ($R^2 = .50$; $F(3, 108)=36.45$; $p < .000$) by the direct measures of the three main components ('attitude', 'social pressure' and 'perceived control') for the near future (for more details, refer to Appendix E).

'squeezed out' the role of 'perceived control'. Therefore, the contribution of 'perceived control' to explaining the willingness to share, both, in the present and in the near future, cannot be confirmed when controlling for 'attitude' and 'social pressure'. However, because there is indication of significance in the simple relation between intention and 'perceived control', it is not possible to infer that 'perceived control' does not have any effect at all. Indeed, the TPB postulates that 'perceived control' can have a direct influence on the actual behaviour - in this case spatial data sharing - and that this link is not mediated by the willingness to engage in spatial data sharing (see Figure 7.1).[42]

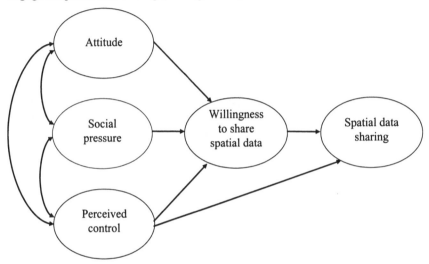

Figure 7.1   Basic model of spatial data sharing
Source: based on Ajzen (1991), repeated from Chapter 5

Ajzen (1991) argued that the relative importance of 'attitude', 'social pressure' and 'perceived control' in the prediction of intention is likely to vary depending on the behaviour that the TPB is applied to. The relative importance of the confirmed contributions of 'attitude' and 'social pressure' at present is indicated by their standardised regression coefficients (Beta; see Table 7.2). These show that the contribution of 'perceived social pressure' (Beta = .462) to explaining willingness to share is almost twice as important as the contribution of 'attitude' (Beta = .255).

It can be concluded, therefore, that at the level of the main components, the 'social pressure' perceived from important referents to engage or not to engage in spatial data sharing is the most important determinant of the current willingness to

---

[42] Due to the constraints detailed in Chapter 5, the measures of behaviour in this research were used to assess *past* behaviour. However, the link between 'perceived control' and behaviour suggested by the TPB is relevant only for measures of behaviour taken *subsequently* to assessing intentions and the three main components ('attitude', 'social pressure', and 'perceived control'). Hence, the extent to which 'perceived control' over spatial data sharing has an influence on actual behaviour is not examined.

share spatial data at present. For the near future, the confirmed contributions of 'attitude' and 'social pressure' are almost equal in terms of relative importance (Beta = .324 and Beta = .434, respectively; see Table 7.7) in explaining the willingness to engage in spatial data in the near future.

Determinants of Attitude towards Spatial Data Sharing

The 'attitude' towards spatial data sharing was expected to be indicated by the results of the survey for the four domains of behavioural beliefs about resources outcomes, outcomes for organisational activities, strategic position outcomes, and social outcomes. As each of the four domains consists of one or more scales, the combined beliefs (*be*) belonging to a particular scale were summed to form an index ($\sum be$). The resultant indices are used as the independent variables for the multiple regression analysis of the determinants of the 'attitude' towards spatial data sharing (see Table 7.4; detailed results are presented in Appendix E). The results indicate that this set of independent variables accounts for 34% of the variation in the 'attitude' of organisations towards spatial data sharing (R^2= .34, F(8, 103) = 6.55, p< .000).

The significance levels of the regression coefficients of the independent variables show that only the scales for knowledge creation (p= .012) and social outcomes (p= .023) seem to serve as significant predictors of 'attitude'. The relative importance of their contributions to the prediction of 'attitude' is almost equal, as indicated by the standardised regression coefficient (Beta = .279 and Beta = .248, respectively).

Table 7.4 Standardised regression coefficients with attitude as dependent variable

Independent variables		Standardised regression coefficient (Beta)	Significance *p*
1. Resource outcomes			
Costs	$\sum_{i=1}^{5} b_i e_i$	-.061	.470
Benefits	$\sum_{i=6}^{10} b_i e_i$	.190	.053
Spatial data outcomes	$\sum_{i=11}^{14} b_i e_i$	.102	.356
2. Organisational activities	$\sum_{i=15}^{17} b_i e_i$	.030	.790
3. Strategic position outcomes			
Loss of control of spatial data	$\sum_{i=18}^{21} b_i e_i$	.065	.482
Knowledge creation	$\sum_{i=22}^{25} b_i e_i$	.279	**.012**
Inter-organisational relations	$\sum_{i=26}^{27} b_i e_i$	-.170	.122
4. Social outcomes	$\sum_{i=28}^{29} b_i e_i$	.248	**.023**

Determinants of Perceived Social Pressure

The 'social pressure' to engage or not to engage in spatial data sharing was expected to be indicated by the results of the survey for the five different domains of normative beliefs about the expectations of important referents (GIS community, market, institutional, and organisational pressure, and moral norms). The products (*nm*) that were made up of each normative belief and the respective 'motivation to comply' were summed within each domain (Σnm) to provide the independent variable for the multiple regression analysis of the determinants of 'social pressure' to engage in spatial data sharing (see Table 7.5, detailed results are presented in Appendix E). The results of the regression analysis show that almost 40% of the variance in the perceived 'social pressure' to share spatial data can be explained (R^2= .398, F(5, 106)= 14.03, p < .000).

Table 7.5   **Standardised regression coefficients with social pressure as dependent variable**

Independent variables		Standardised regression coefficient (Beta)	Significance p
1. GIS community pressure	$\sum_{i=1}^{7} n_i m_i$	.236	**.035**
2. Market pressure	$\sum_{i=8}^{11} n_i m_i$	.063	.542
3. Institutional pressure	$\sum_{i=12}^{14} n_i m_i$	.101	.362
4. Organisational pressure	$\sum_{i=15}^{18} n_i m_i$	.252	**.012**
5. Moral norm pressure	$\sum_{i=19}^{20} n_i m_i$	.134	.161

The significance levels of the regression coefficients indicate that only the domains of GIS community and organisational pressures make significant contributions to the prediction of 'social pressure' (see Table 7.5). The relative importance of their contributions to the prediction is about the same (Beta = .236 and Beta = .252, respectively).

It is worth noting that the contribution of institutional pressures to the prediction of 'social pressure' cannot be confirmed. This presents an interesting finding since one of three institutional referents in this domain is the National Spatial Information Framework (NSIF) directorate in South Africa. One might have expected that the NSIF would emerge as an important referent for organisations with respect to their engagement in spatial data sharing.

The NSIF directorate is situated in a national government department, the Department of Land Affairs. Examination of the correlation between the questions referring to GIS national government departments within the GIS community and the specific items referring to the NSIF reveals that the correlations are significant (see Appendix F). This suggests that responses to the questions about the GIS national government departments are highly related to those about the NSIF. Further support for this relationship is provided by the results of factor analysis of these items[43] which shows that they are tapping the same underlying concept (see Appendix F). This is not to suggest that responses to GIS national government departments were exclusively interpreted by the respondents as referring to the NSIF. Rather, these results give weight to the argument that, according the perception of the respondents, the NSIF is conceptually linked to the GIS national government departments, thereby indirectly including the NSIF in the referents from the GIS community. Yet the role of the NSIF as an important referent for organisations regarding their engagement in spatial data sharing could not be *directly* confirmed in the domain of institutional pressures.

Determinants of Perceived Control

'Perceived control' over spatial data sharing was expected to consist of six domains of control beliefs: technical and interpersonal skills, control over resources, spatial data position, finding sharing partners, and opportunities for spatial data sharing. Each of the six domains consists of one or more scales. The combined control beliefs (*cp*) belonging to a particular scale were summed to form an index (Σcp). These indices serve as the independent variables for the multiple regression analysis of the determinants of the 'perceived control' over spatial data sharing (see Table 7.6; detailed results are presented in Appendix E). The results of the analysis indicate that 32% of the variance in 'perceived control' can be explained ($R^2 = .324$, $F(16, 95) = 2.851$, $p = .001$).

---

[43] In this case, factor analysis was carried out with the 'uncombined' beliefs, i.e. using the separate normative beliefs and 'motivation to comply' with each referent, the GIS national government departments and the NSIF. In contrast, the factor analysis reported in Appendix E was carried out on the combined beliefs, i.e. the products resulting from combining the normative beliefs with the willingness to comply.

Table 7.6   Standardised regression coefficients with perceived control as dependent variable

Independent variables		Standardised regression coefficient (Beta)	Significance p
1. Technical skills			
Spatial data skills	$\sum_{i=1}^{5} c_i p$	-.038	.720
Metadata skills	$\sum_{i=6}^{10} c_i p$	.301	**.011**
IT skills	$\sum_{i=11}^{14} c_i p$	.022	.854
2. Interpersonal skills			
Networking and collaboration skills	$\sum_{i=15}^{19} c_i p$	-.094	.444
Negotiation skills	$\sum_{i=20}^{23} c_i p$	.256	**.015**
Past experience	$\sum_{i=24}^{26} c_i p$	-.082	-.815
3. Resource control	$\sum_{i=27}^{29} c_i p$	.173	.126
4. Spatial data position			
a) Control over internal spatial data			
Importance of internal spatial to other organisations	$\sum_{i=30}^{30} c_i p$	.072	.516

Table 7.6   continued

Independent variables		Standardised regression coefficient (Beta)	Significance p
Control aspects	$\sum_{i=31}^{34} c_i p$	.-474	**.000**
Availability and stability of alternative spatial data sources	$\sum_{i=35}^{36} c_i p$	-.010	.931
b) Dependence on external spatial data			
Importance of external spatial data to own organisation	$\sum_{i=37}^{37} c_i p$	.237	**.030**
Dependence aspects	$\sum_{i=38}^{41} c_i p$	-.083	.406
Availability and stability of alternative spatial data sources	$\sum_{i=42}^{44} c_i p$	-.068	.562
Self-sufficiency	$\sum_{i=45}^{46} c_i p$	.015	.898
5. Finding sharing partners	$\sum_{i=47}^{50} c_i p$	-.093	.351
6. Opportunities	$\sum_{i=51}^{56} c_i p$	-.004	.969

The significance levels of the regression coefficients of the scale indices indicate that significant contributions to the prediction of 'perceived control' were made by metadata skills, negotiation skills, control aspects, and the importance of external spatial data to an organisation. The most important contribution was made by the control aspects (Beta = -.474), followed by the beliefs about metadata skills (Beta = .301). Finally, equal contributions were made by the beliefs about negotiations skills (Beta = .256) and the belief about the importance of external spatial data (Beta = .237).[44]

Summary of Determinants

The foregoing analysis of the specific determinants of the perceived willingness to share spatial data has shown that a small number of contributing factors can be identified as being significant predictors of the willingness to share spatial data. Figure 7.2 depicts the model of spatial data sharing where these contributing factors have been highlighted.

At the level of the three main components, 'social pressure' was the most important factor, followed by 'attitude' towards sharing in determining the current willingness to share. The importance of 'perceived control' could not be confirmed. These results suggest the important role of outcome considerations of sharing ('attitude') and expectations by important referents ('social pressure') in influencing the intention to engage in spatial data sharing. At the same time, the importance of the technical aspects of sharing in terms of the skills and resources considered under 'perceived control' is somewhat diminished.

At the level of the indices, knowledge creation and social outcomes were most important in predicting the 'attitude' towards spatial data sharing, and GIS community and organisational pressure were most important in contributing to 'social pressure'.

An additional finding was that the role of NSIF as an important referent for organisations with respect to their engagement in spatial data sharing was not manifested directly.

---

[44] The negative beta coefficient of the scale index for control aspects means that, on average, for each additional score on the index of beliefs about control aspects, the score for 'perceived control' is reduced by almost half a point on the seven-point scale

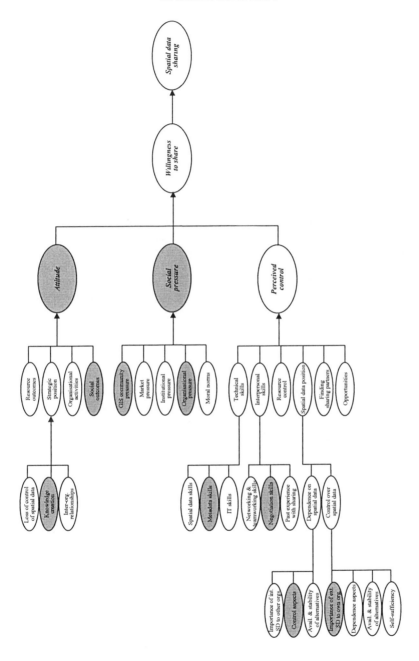

Figure 7.2 Summary of scale index determinants

Fostering a Spatial Data Sharing Culture

The analysis up to this point has suggested that 'social pressure' and 'attitude' influence the willingness to share spatial data. As was presented in Chapter 6, the willingness of individuals in organisations to engage in spatial data sharing was low and the majority of the respondents was undecided as to whether their organisation should or should not engage in spatial data sharing activities. The empirical evidence also showed that although some sharing activities are actually taking place, the scope and scale of these exchanges was still very limited at the time of the survey. It is, therefore, of interest not only to predict and understand the particular behaviour under consideration - spatial data sharing across organisational boundaries - but also to examine options to influence the likelihood of its occurrence to foster a 'spatial data sharing culture'.

For this part of the analysis, the respondents are divided into three groups according to their responses regarding their organisation's willingness to engage in spatial data sharing (*willing, undecided,* and *unwilling*). Examination of the beliefs that have been identified as influencing 'social pressure' and 'attitude' for significant differences between the means of these three groups can provide insight into how the beliefs may need to be addressed in order to increase the willingness to share spatial data (for detailed results, see Appendix G).

Referents Inside and Outside: Who Wants Organisations to Share?

'Social pressure' does appear to positively influence the willingness to share spatial data. The analysis of the determinants of 'social pressure' has identified two important sets of referents for organisations with respect to spatial data sharing; firstly, the GIS community, and, secondly, actors inside the organisation. With respect to the GIS community, Figure 7.3 shows the mean score on the normative beliefs (perceived pressure) indicating the expectations of each sector within the GIS community as perceived by each of the intention groups.

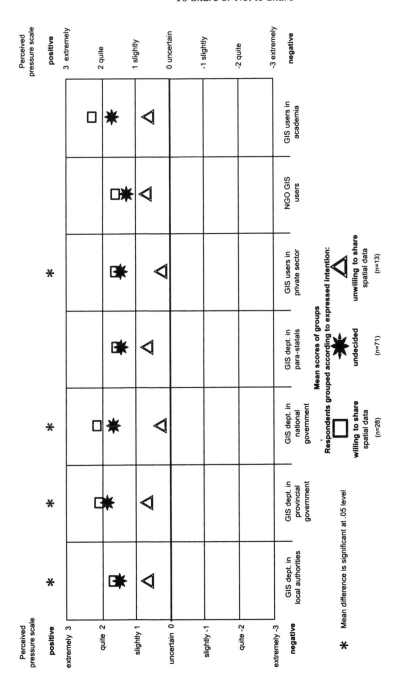

Figure 7.3   Perceived GIS community pressure by intention group

This figure should be read as follows. Each column refers to a separate belief (see labels in bottom row). For each belief, the mean score for each of the three groups (*willing, undecided,* and *unwilling*) is displayed. Significant differences between the means of these three groups are indicated by a '*' above the column.

The mean scores show that the referents from the different sectors in the GIS community were perceived to be in favour of their organisations' engagement in sharing spatial data. However, comparison between the scores for the three intention groups yields a more refined result. The pressure perceived from the GIS community was consistently high (around 'quite positive') for those respondents willing to share but the perceived pressure by those respondents who indicated their organisation to be unwilling to engage in spatial data sharing was consistently much lower (below 'slightly positive'). Significant differences in the mean scores of the three intention groups are evident for the pressure perceived from GIS departments in local, provincial and national government and in the private sector. This comparison suggests that the pressure perceived from the different GIS community sectors is not as strong and encouraging as it might be. The stimuli that organisations are receiving to share could be clearer and more encouraging.

The extent to which organisations, as indicated by their respondents, are motivated to comply in general with the expectations of the referent groups from the GIS community is indicated by the mean scores in Figure 7.4. For most of the GIS community sectors, the 'motivation to comply' with them differs significantly in magnitude across the three groups of respondents (willing, undecided and unwilling). Although the mean scores for the 'undecided' intention group were somewhat lower, the pattern of these scores paralleled those of the 'willing' group. Both groups were 'quite' motivated to comply with GIS departments in provincial and national government but less so with the remaining sectors in the GIS community. Organisations perceived to be unwilling to share by their representatives are only 'moderately' motivated to comply with the different GIS community sectors.

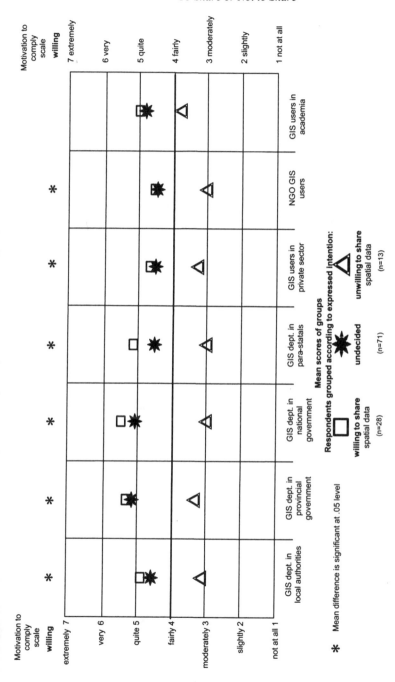

Figure 7.4 'Motivation to comply' with GIS community by intention group

The overriding goal of this study is to consider how spatial data sharing can be fostered. The influence of 'social pressure' on the willingness to share has already been established in the foregoing analysis. The results of the 'willing' intention group show that high perceived pressure matched by high scores on the 'motivation to comply' can contribute to a high level of 'social pressure' and, in turn, to a high willingness to share.

The second source of 'social pressure' was the important referents for sharing inside the organisation. The average scores on normative beliefs (perceived pressure) for each of the three intention groups are shown in Figure 7.5. This indicates significant differences as to how the three groups perceived the expectations (with respect to spatial data sharing) of other departments within the organisation, of the management of the organisation, and of 'champions' for spatial data sharing in the organisation. The 'willing' group consistently perceived these referents to be in favour of sharing, with scores above 'quite positive'; the 'undecided' group perceived the expectations only 'slightly to quite' positive while the 'unwilling' group indicated only barely positive favourable expectations ('uncertain to slightly' positive) of the organisational referents.

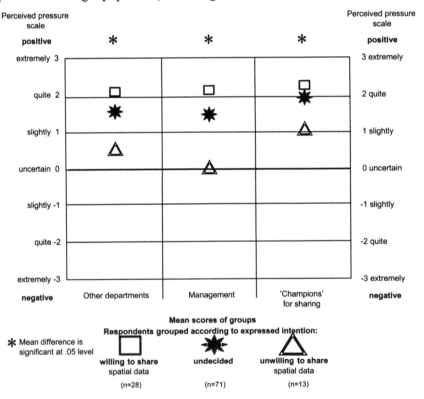

Figure 7.5   Perceived organisational pressure by intention group

The mean scores for the 'undecided' and the 'unwilling' group indicate that although the respondents perceived some positive stimulus for sharing from the referents inside their organisation, this encouragement was not perceived to be very strong. While it is important to take into account that this is the perspective of the respondents, it does suggest that the encouragement from referents inside the organisation to engage in spatial data sharing could to be substantiated.

In this respect, one of the key referents inside the organisation that can be targeted is the management of GIS-using organisations. Since much of the existing NSIF initiative is focused at the technical level to establish standards, procedures and clearinghouses, the individuals addressed by, and involved in, the NSIF within the organisations are at a technical level. A frequent complaint from these GIS personnel is that, because their management has no understanding or appreciation of GIS, these systems are not valued and requirements, such as the availability of relevant spatial data sets, are not recognised. The NSIF initiative could target the management of GIS-using organisations to increase their awareness and understanding of spatial data sharing activities for using their organisation's GIS more effectively so that they would signal their support for, and actively encourage, such activities. This may be a way of enhancing the role of the management as an important referent within the organisation with respect to spatial data sharing.

The average scores on the 'motivation to comply' more generally with the referents inside the organisation, including the organisational goals or mission, for each of the three intention groups are shown in Figure 7.6.

For all three groups, the 'motivation to comply' was above 'fairly willing' but the mean scores for the 'undecided' and 'unwilling' groups were consistently lower than those of the 'willing' group. This difference is statistically significant only for the 'motivation to comply' with 'champions' for spatial data sharing in the organisation.

High scores on the normative beliefs about organisational pressure, coupled with high scores on the 'motivation to comply' with the organisational referents, are conceived to contribute to a high level of 'social pressure', which, in turn, would contribute to a high willingness to share. Since the 'motivation to comply' with the referents inside the organisation, as well as with those outside (the GIS community sectors), was low for the 'undecided' and 'unwilling' groups, specific steps are considered in the next section on how to address the low 'motivation to comply'.

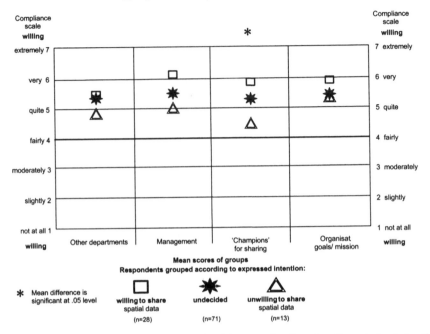

Figure 7.6   **'Motivation to comply' with organisational referents by intention group**

Changing the 'motivation to comply' with important referents   According to the TPB, beliefs are the basic determinants for any behaviour and, therefore, change in behaviour is brought about by effecting changes in beliefs. Although the literature on persuasive communication research argues that a diverse range of variables (source, message, recipient and context) affect the success of attitude and belief change, the focus here is less on the psychological mechanisms of persuasive communication that influence the tendency of people to accept information and how they integrate information.[45] Rather, in this section, steps that need to be taken in the policy arena are considered.

The understanding of the factors that influence the behaviour can provide guidelines for the formulation of persuasive communication that will be effective in *changing* behaviour. The beliefs that were identified in the preceding analysis can serve as the basic argument for persuasive communication; alternatively, beliefs may be changed indirectly by presenting other arguments known to be related to the beliefs (Fishbein and Ajzen, 1980).

As suggested above, there may be indirect means by which beliefs can be changed. Since it is difficult to foresee direct ways of influencing the motivation of organisations to comply with the GIS community and with the organisational referents, indirect means may be appropriate in order to address and increase the

---

[45] For a detailed overview of research on attitude and belief change, see Petty and Wegener (1998) and Eagly and Chaiken (1993).

'motivation to comply'. To this end, it is considered whether a relationship between the 'motivation to comply' with particular referents and the evaluation of the possible outcomes of spatial data sharing exists. Such a relationship would imply that the more positively the consequences of engaging in spatial data sharing are evaluated, the greater the 'motivation to comply' with the expectations of the GIS.

Specifically, positive correlation between 'motivation to comply' and particular 'outcome evaluations' is expected to indicate that, by using persuasive communication to create awareness about these outcomes, the underlying beliefs about the outcomes may be influenced. This, in turn, can indirectly affect the 'motivation to comply' with the referents and, ultimately, modify 'perceived social pressure' which has been demonstrated to have a strong influence on the willingness to engage in spatial data sharing, as established in the preceding analysis (see section on *Determinants of intention* in this chapter).

A number of significant relationships exists between the 'motivation to comply' with particular sectors in the GIS community and referents inside the organisation, and specific outcomes of spatial data sharing (for detailed results, see Appendix H). This subset of outcome evaluations is examined for the extent to which the evaluations vary across the three intention groups (see Table 7.7). The specific consequences for which significant differences in the mean for the three groups have been established are discussed below.

The mean scores for the outcomes of spatial data sharing for organisational activities in terms of being able to focus on the organisation's core activity, the usefulness of the organisation's GIS, and the quality of decision making in the organisation differ significantly across the three intention groups. The mean scores of the 'unwilling' and 'undecided' groups are considerably lower than those of the 'willing' group. This suggests that one way to influence the 'motivation to comply' would entail awareness creation activities about the potential benefits of spatial data sharing for an organisation's activities. Generally, the publicity output of the NSIF in South Africa makes reference to the benefits of establishing the framework and of spatial data sharing. However, in order to influence the motivation of key individuals in organisations to engage in spatial data sharing, communication about the benefits of spatial data sharing for organisational activities would need to be articulated by specifying explicitly *how* organisations that engage in spatial data sharing are able to benefit by focusing on their core activity, to increase the usefulness of their GIS and to improve the quality of their decision making. In addition, the benefits in terms of savings on administrative efforts that would otherwise be needed for data capture could usefully be mentioned.

Table 7.7   Outcome evaluations by intention group

Outcome evaluations (n = 112)	Mean score Intention group:		
	willing	*undecided*	*unwilling*
Resource outcomes			
Magnitude of savings implications			
Time saved otherwise needed to capture spatial data in-house	5.61	5.51	5.46
Trained staff saved otherwise needed for capturing and maintaining spatial data	5.50	5.20	4.31
Administrative efforts saved otherwise needed for data capture*	**4.89**	**4.90**	**3.69**
Outcomes for spatial data			
Increased quantity of spatial data available to the organisation	6.21	6.06	6.00
Introduction of standards agreed with other organisations	6.21	5.99	5.38
Organisational activities			
Focus on core activity*	**5.46**	**5.39**	**4.54**
Usefulness of the GIS*	**6.29**	**6.14**	**5.54**
Quality of decision making*	**6.00**	**5.94**	**5.31**
Strategic position outcomes			
Loss of control over spatial data			
Unregulated access to organisation's spatial data by other organisations*	**3.82**	**3.46**	**1.85**
Use of organisation's spatial data by organisations other than the sharing partners*	**4.11**	**3.92**	**2.46**
No enforcement of any rules regarding the ownership of, access to and use of the organisation's spatial data*	**3.36**	**3.35**	**2.00**
Knowledge creation outcomes			
Combination of new and existing spatial data sets*	**6.04**	**6.01**	**4.92**
Access to ideas	5.68	5.54	4.77
Inter-organisational relations outcomes			
Increased interdependence	5.25	4.97	4.15
Redistribution of influence	5.29	5.07	4.38
Social outcomes			
Integrated development planning*	**6.29**	**5.59**	**4.31**
Benefits to society at large*	**6.46**	**5.96**	**5.00**

Note: * Differences between means are significant at the 0.05 level (2-tailed) (for detailed test results, see Appendix I)

Responses have been rescaled to 1 (extremely negative connotation) to 7 (extremely positive connotation).

For the items under social outcomes, respondents were not asked to *evaluate* them but only to assess the *likelihood*.

The perceptions of the consequences of spatial data sharing for an organisation's strategic position in terms of loss of control over spatial data also differ significantly across the three groups. The extremely low mean scores of the 'unwilling' group indicate that unregulated access, use of an organisation's data by others and lack of enforcement of rules regarding spatial data are perceived by the respondents as strong threats to their organisation's strategic position. While much of the promotion of the NSIF stresses benefits, it would be helpful to reduce the fears associated with the negative consequences of spatial data sharing as opposed to solely emphasising positive outcomes. In order to do this, the threats to an organisation's strategic position would need to be addressed. Specifically, how unregulated access to an organisation's spatial data by other organisations can be limited and how the use of an organisation's spatial data can be restricted to the specific organisations that are considered sharing partners. In addition, it would be important to clarify the mechanisms, if any, that can be put in place to ensure that rules regarding the ownership of, access to and use of the organisation's spatial data can be enforced. Addressing the issues that are perceived as a threat to the strategic position of the respondents' organisations, may be expected to increase the 'motivation to comply' with the important referents for spatial data sharing.

Another aspect of spatial data sharing outcomes for an organisation's strategic position, namely knowledge creation outcomes, is related to the 'motivation to comply' with important referents. In particular, the mean scores of one mechanism of knowledge creation, the combination of new and existing spatial data sets from different sources, differ significantly for the three groups. Even so, this outcome is generally perceived as an opportunity rather than as a threat. Reference to, and awareness creation of, this intangible outcome of spatial data can be expected to increase the motivation to comply with important referents.

Finally, the perception of social outcomes of spatial data sharing differs significantly across the three groups but, even for the 'unwilling' group, the mean scores are reasonably high ($m = 4.31$ and $m = 5.00$, respectively). These results suggest that integrated development planning can present an incentive for spatial data sharing. While general benefits to society at large are often mentioned in the existing promotion of spatial data sharing activities in South Africa, specific reference to integrated development planning has not been made. Specific mention of integrated development planning in the context of promoting spatial data sharing may be expected to influence the 'motivation to comply'.

Opportunities or Threats: To Share or Not to Share?

The 'attitude' towards spatial data sharing has an influence on the willingness of organisations to engage in spatial data sharing. The two important sources of 'attitude' that were identified are beliefs about knowledge creation outcomes and beliefs about social outcomes of spatial data sharing.

The items in the knowledge creation domain captured the beliefs about the intangible consequences of sharing. These intangible outcomes of spatial data sharing were generally perceived as opportunities rather than as threats (see mean scores in Figure 7.7).

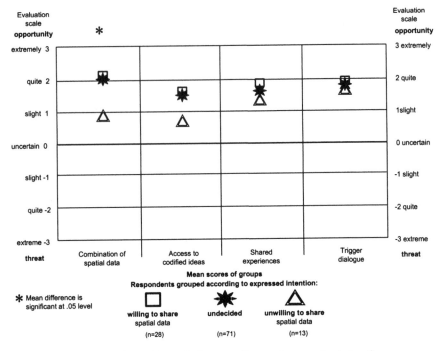

Figure 7.7   Evaluation of knowledge creation outcomes by intention

However, comparison of the scores for the three intention groups reveals considerable discrepancy. With respect to the evaluation of the combination of spatial data, the differences between the mean scores of the three intention groups are significant. Although all four knowledge creation outcomes were generally perceived as opportunities, the mean scores of the 'undecided' and the 'unwilling' group were consistently lower than the score of the 'willing' group (below 'quite an opportunity' or even only 'slight opportunity').

The preceding analysis has already established the influence of 'attitude' on the willingness to share. Hence, it can be argued that high scores on the outcome evaluation, coupled with high scores on the behavioural beliefs (likelihood), would contribute to a very positive 'attitude' towards spatial data sharing, which, in turn, would contribute to a high willingness to share. The mean scores for the behavioural beliefs (likelihood) show that the 'undecided' and the 'unwilling' groups perceived the knowledge creation consequences to be only 'slightly' to 'quite likely' to result from spatial data sharing (see mean scores in Figure 7.8) whereas the mean scores of the 'willing' group were consistently higher. The difference between the mean scores of the three groups was significant for the likelihood that spatial data sharing would trigger dialogue and collective reflection across organisational boundaries.

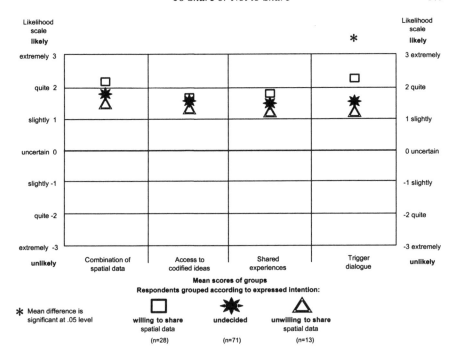

Figure 7.8   Likelihood of knowledge creation outcomes by intention

While the analysis has shown that these knowledge creation 'moments' present incentives for spatial data sharing, many respondents did not consider them to be very positive or likely. The way in which this might be addressed is by providing information about how organisations can benefit from spatial data sharing in terms of these intangible outcomes with examples that are salient to them. Specifically, reference could be made to the four dimensions of knowledge creation by detailing how organisations can benefit from spatial data sharing because it may: 1) trigger dialogue and collective reflection across organisational boundaries, such as joint problem solving, and involve articulating and comparing spatial data-related concepts; 2) allow access to ideas that are codified in spatial data and enable their translation, combination and application in different organisations; 3) provide the means to observe spatial data skills in personal interaction and to create shared mental models and experiences across organisational boundaries; and 4) enable the combination of new and existing spatial data sets from different organisations to create new knowledge. Detailed examples can serve to influence both how positive and how likely these outcomes are considered by the target audience.

The second group of beliefs influencing 'attitude' included beliefs about the social outcomes of sharing (shown in Figure 7.9).

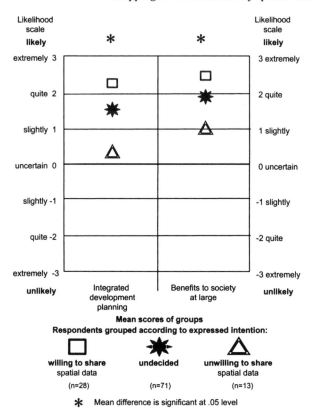

Figure 7.9   Likelihood of social outcomes by intention

Since these outcomes - integrated development planning and benefits to society at large - were regarded as being inherently positive, respondents were not asked to evaluate these outcomes. Only the likelihood of these outcomes was assessed. Comparison of the mean scores for the three intention groups shows that there are significant differences. The scores of the 'willing' group were above 'quite' likely while those of the undecided group were 'slightly' to 'quite likely' and of the 'unwilling' group 'slightly likely' or below. This suggests that high scores for the likelihood that sharing would result in these social outcomes contribute to a positive 'attitude' towards spatial data sharing, which, in turn, contributes to a high willingness to share.

As was mentioned in the section *Changing the 'motivation to comply' with important referents* in this chapter, integrated development planning presents an incentive for spatial data sharing that has not been previously linked to spatial data sharing in South Africa. In order to foster a spatial data sharing 'culture', this is another aspect that could be included in addressing the target audience, i.e. the GIS community.

Summary

This section has examined the scores on the determinants of 'social pressure' and 'attitude' in order to gain an understanding of the beliefs that may need to be addressed in order to foster the willingness to engage in spatial data sharing. Considerable differences have become apparent between the scores of the respondents (grouped according to their intention into 'willing', 'undecided' and 'unwilling') for the normative and behavioural beliefs, the motivation to comply and outcome evaluations. Specifically, this analysis has revealed that:

- those respondents who were undecided or unwilling to engage in spatial data sharing indicated that the expectations from referents inside and outside the organisation to share were not perceived to be as strong and as encouraging as they might be. The stimuli that organisations are receiving from the GIS community to share could be clearer and more encouraging, indicating the current lack of a spatial data sharing 'culture'. As a key referent, the management of GIS-using organisations could be targeted by the spatial data infrastructure initiative to increase their awareness and understanding of spatial data sharing activities for the effectiveness of their organisation's GIS, so that they provide support and encouragement for such activities;

- the motivation to comply with the referents inside and outside the organisation was low. While it is important to specify particular benefits of engaging in spatial data sharing, such as how to benefit in terms of organisational activities, the disincentives for spatial data sharing also have to be addressed. In particular, the importance of reducing the fears associated with losing control over spatial data has been highlighted;

- the knowledge creation aspects of spatial data sharing are perceived as incentives but they need to be referred to in more detail;

- the government initiative, 'integrated development planning', has emerged as another incentive for spatial data sharing; therefore, specific reference to integrated development planning should be made in the context of the spatial data infrastructure initiative in South Africa to contribute to a positive attitude towards spatial data sharing.

Conclusions

This chapter has shown how the behavioural approach developed in this study can be used to understand decisions by individuals in organisations to engage in spatial data sharing across organisational boundaries. It has verified that the model of the willingness of organisations to share spatial data developed in Chapter 5 is valid and reliable.

Furthermore, the specific determinants of the willingness to share spatial data have been identified. The chapter has shown that the intention of key individuals within organisations to engage in spatial data sharing can be predicted from corresponding measures of 'attitude' towards spatial data sharing and 'social pressure' from important referents to engage in spatial data sharing, although somewhat greater emphasis was placed on the 'social pressure' component. The results show that the technical aspects of sharing in terms of the skills and resources considered under 'perceived control' could not be confirmed as an important factor in influencing the willingness of individuals in organisations to engage in spatial data sharing.

It has also been demonstrated that the 'attitude' towards spatial data sharing is influenced by behavioural beliefs about, and outcome evaluations of, knowledge creation and social outcomes and that 'social pressure' is a function of normative beliefs about, and the 'motivations to comply' with, the GIS community and with referents inside the organisation. Finally, a more detailed and substantive explanation of spatial data sharing across organisational boundaries was obtained by examining the actual scores on the identified determinants of 'attitude' and 'social pressure' according to three distinct groups of respondents - those who perceived their organisations to be *willing* to engage in spatial data sharing, those who perceived their organisations to be *undecided* and those who perceived their organisations to be *unwilling*.

An additional finding is that the NSIF directorate did not emerge as an important referent for organisations with respect to their engagement in spatial data sharing. The results suggest that, according the perception of the respondents, the NSIF is conceptually linked to the GIS national government departments, thereby indirectly including the NSIF in the referents from the GIS community. Yet the role of the NSIF as an important referent for organisations regarding their engagement in spatial data sharing could not be *directly* confirmed in the domain of institutional pressures.

In the final chapter, conclusions are drawn regarding the usefulness and applicability of the model, to broader theoretical considerations and for policy.

Chapter 8

Conclusions

Introduction

This research has examined the propensity of organisations to engage in spatial data sharing across organisational boundaries, focusing on individual decision makers within different organisations and taking South Africa as a case study. The organising framework was the TPB, developed by Ajzen (1991). This theory was used to integrate qualitative empirical evidence with insights drawn from several bodies of literature in a systematic fashion to arrive at a model of the willingness of organisations to engage in spatial data sharing. The model developed in this research was applied to assess the factors that are likely to influence the willingness of key individuals embedded in organisations to engage in spatial data sharing across organisational boundaries.

The validation of the model has demonstrated that it can be used to address the research questions regarding the determinants of the willingness of organisations to engage in spatial data sharing. The main research findings based on the analysis of the survey data indicated that, overall, the willingness of the organisations in the sample to share spatial data across organisational boundaries was low, with a slight increase expected in the near future. Self-reports of actual sharing activities revealed that, although most respondents indicated some involvement of their organisations in spatial data sharing across organisational boundaries, the degree of spatial data sharing activities was still very limited. Furthermore, the research provided a basis for identifying the specific determinants of the willingness to share spatial data across organisational boundaries in South Africa.

The purpose of this concluding chapter is to discuss the usefulness and applicability of the model in the light of broader theoretical considerations and for policy. The chapter is organised in five parts. In the first section, the research questions are re-stated and discussed in the light of the findings. Next, conclusions are drawn about the contribution of the model developed here to the understanding of the diffusion of the GIS technological innovation and the extent to which it offers a means of examining the social or behavioural aspects of the diffusion process more generally in other contexts of GIS usage - and more broadly still - in studies of the diffusion of innovations in the science and technology policy research field. The third section reflects on the specific methods used to implement this theoretical approach and considers alternative approaches. Section four points towards new areas of research and section five concludes with some final remarks.

Research Questions and Results

This research has explored the determinants of the willingness of organisations to engage in spatial data sharing across organisational boundaries. The aim was to examine how spatial data sharing among the different actors involved with spatial data can be fostered in order to extend the availability of, and access to, spatial data so that GIS may be used more effectively. To address specific aspects of this principal question, three more specific questions were posed for the empirical enquiry:

1. How willing or resistant are key individuals within different organisations to engage in spatial data sharing across organisational boundaries?
2. What is the extent of spatial data sharing activities already taking place?
3. What are the factors that determine the willingness or resistance of key individuals within their organisations to engage in spatial data sharing?

Spatial data sharing was defined in this study to involve exchange relationships that may or may not include financial payment or payment in kind and that entail making spatial data accessible to, or from, other parties under certain terms and conditions. The constructs systematised in this research provided a framework within which these questions could be addressed. The research generated empirical evidence to aid in examining the incentives and disincentives for organisations to engage in spatial data sharing. The following discussion of the empirical results considers the similarities and differences between the insights generated by the research and elements of the practices and incentives created by the current spatial data sharing initiative by the policy and organisational set-up in South Africa.

Who is Willing to Share?

The first research question addressed the willingness of key individuals within organisations to engage in spatial data sharing. In order to explore this question, the intention concept of the TPB was used to assess the willingness of organisations in South Africa, as perceived by key individuals within the organisation, to engage (or not) in spatial data sharing activities across organisational boundaries.

The results suggest that, contrary to the underlying assumption in the design of, and in many discussions on, spatial data infrastructures, in South Africa, the willingness of different organisations to engage in spatial data sharing cannot be taken for granted. The finding of this research is that the actual motivation of organisations in the sample was not in favour of participation in spatial data sharing. Overall, the willingness of the representatives of organisations in the sample to share spatial data across organisational boundaries was low, with only a slight increase expected in the near future. The majority of the respondents reported that their organisation was undecided about whether to engage in spatial data sharing and a small share indicated that their organisations were unwilling to share spatial data with other organisations.

The results of the analysis further indicate that the willingness of organisations, as expressed by the respondents in the sample, to engage in spatial data sharing did not vary significantly for the different sectors in the GIS community. There were also no significant differences in the willingness to share according to the perceived extent of self-sufficiency or dependence of organisations on spatial data. Contrary to what was expected, the willingness to share did not differ with respect to the organisations' spatial data position, i.e. whether organisations were predominantly giving spatial data to, or receiving spatial data from, other organisations.

Who IS Sharing?

The second research question addressed the extent of actual spatial data sharing behaviour. It asked to what extent spatial data sharing activities were already taking place. Self-reports of actual sharing activities that were collected during the survey revealed that, although most respondents indicated some involvement of their organisations in spatial data sharing across organisational boundaries, the degree of spatial data sharing activities was still very limited. The scope of spatial data sharing among organisations can be categorised according to the following levels (Obermeyer and Pinto, 1994; Pinto and Onsrud, 1995): a) situation specific, project driven, non-routine, and non-recurring, to solve specific problems; b) regular sharing and exchange of information using protocols and procedures for limited sharing parties; and c) routine sharing, standardised, generalisable patterns of exchange, accessible to all parties in terms of location and format. The results indicate that only the first level of ad hoc, non-routine sharing had been reached by those included in the sample at the time of the survey. This suggests that, in accordance with the objectives of the spatial data sharing initiative in South Africa, the NSIF, there is still much room for increasing spatial data sharing activities across organisational boundaries.

Incentives and Disincentives for Spatial Data Sharing in South Africa - Promoting the Benefits and Addressing the Fears

The third research question addressed the determinants of the perception of key individuals within organisations that influence their willingness or resistance to engage in spatial data sharing across organisational boundaries. Based on the model developed in this research, it was possible to identify the incentives and disincentives for key individuals within organisations to share spatial data. These findings form a profile of perceptions that provides a basis for addressing the issue of spatial data sharing more effectively at the policy making level.

The results of the statistical analysis show that the willingness of key individuals within organisations to engage in spatial data sharing can be predicted from corresponding measures of 'attitude' towards spatial data sharing and 'social pressure' from important referents to engage in spatial data sharing. Somewhat greater emphasis was placed on the 'social pressure' component. It was found that the technical aspects of sharing in terms of the skills and resources considered under 'perceived control' could not be confirmed as an important factor in influencing the willingness of individuals in organisations to engage in spatial data sharing. This finding suggests that, while the technical aspects that are currently the focus of attention in the context of spatial data sharing initiatives - such as the interoperability of different GIS applications and spatial data sets, the establishment of standards, and the implementation of clearinghouses - are necessary, it cannot be expected that their resolution will be sufficient to overcome the obstacles to spatial data sharing across organisational boundaries.

An additional finding was that the NSIF directorate did not emerge as an important referent for organisations with respect to their engagement in spatial data sharing. As outlined in Chapter 2, the aim of the NSIF is to build a spatial data infrastructure in South Africa in pursuit of social and economic goals by promoting spatial data sharing among different organisations. In the context of diffusion research, the intended role of the NSIF can be conceptualised as that of a change agent. Change agents are seen as opinion leaders who can influence the attitudes and behaviour of individuals with respect to a particular innovation (Rogers, 1995). Since the analysis could not *directly* confirm the NSIF as an important referent for organisations regarding their engagement in spatial data sharing, this finding suggests that the capacity of the NSIF to influence the decisions of key individuals within organisations to engage in spatial data sharing is still limited and that it could to be strengthened. The additional results of the analysis revealed ways in which the effectiveness of the NSIF to encourage greater spatial data sharing could be improved.

A more detailed and substantive explanation of spatial data sharing across organisational boundaries was obtained by distinguishing between three distinct groups of respondents - those who perceived their organisations to be *willing* to engage in spatial data sharing, those who were *undecided* and those who were *unwilling*. This revealed six points that are important to take into account for the current spatial data sharing initiative in South Africa in order to foster a spatial data sharing 'culture'. These are:

1) As a key referent *inside* the organisation for spatial data sharing activities, the management of GIS-using organisations, rather than mainly the technical GIS personnel - as is currently the case, could be targeted by the spatial data infrastructure initiative in order to increase their awareness and understanding of spatial data sharing activities for the effectiveness of their organisation's GIS and so that they could provide support and encouragement for such activities.

2) Although they are key referents *outside* the organisation for spatial data sharing activities, the different sectors of the GIS community do not yet seem to be providing clear and encouraging stimuli for spatial data sharing. The results of the analysis also suggest that, according the perception of the respondents, the NSIF is conceptually linked to the GIS national government departments, thereby indirectly including the NSIF in the referents from the GIS community. Yet the role of the NSIF as an important referent for organisations regarding their engagement in spatial data sharing could not be *directly* confirmed. In combination with the assessment of actual sharing behaviour which revealed that the extent of sharing was still very limited,

these results suggest that a spatial data sharing 'culture' has not yet developed. This adds further weight to the remaining findings of the research that can be used to guide the fostering of such a sharing culture.

3) While it is important to specify particular benefits of engaging in spatial data sharing, the results show that the disincentives for spatial data sharing also need to be addressed. In particular, the importance of reducing the fears associated with losing control over spatial data has been highlighted. While much of the promotion of the NSIF stresses the benefits of spatial data sharing, it would be helpful to reduce the fears associated with the negative consequences of such activities rather than solely emphasising positive outcomes. The threats to an organisation's strategic position would need to be addressed by specifying how unregulated access to an organisation's spatial data by other organisations can be limited and how the use of an organisation's spatial data can be restricted to the particular organisations that are considered sharing partners. It would also be important to clarify the mechanisms that might be put in place to ensure that rules regarding the ownership of, access to, and the use of, the organisation's spatial data can be enforced. Particular benefits of spatial data sharing should refer to the potential benefits of spatial data sharing for an organisation's activities.

4) Generally, the promotion of the NSIF in South Africa makes reference to the benefits of establishing the national spatial information framework and of spatial data sharing. However, the results of this study indicate that it would be beneficial to articulate explicitly how organisations that engage in spatial data sharing are able to benefit by focusing on their core activity, to increase the usefulness of their GIS and to improve the quality of their decision making. Furthermore, it is important to mention the specific benefits of spatial data sharing in terms of the savings in administrative efforts that would otherwise be needed for data capture.

5) Also, aside from the cost-benefit approach usually adopted in the promotion of spatial data sharing initiatives, the results of this research showed that more intangible consequences of spatial data sharing, such as the opportunity to gain new insights, can act as incentives for spatial data sharing. They should be referred to in detail by providing information about how organisations can benefit from spatial data sharing in terms of these intangible outcomes with salient examples. Specific reference could be made to these by detailing how organisations can benefit from spatial data sharing because it may a) trigger dialogue and collective reflection across organisational boundaries, such as joint problem solving, and involve articulation and comparison of spatial data-related concepts, b) allow access to ideas that are codified in spatial data, enabling their translation, combination and application in different organisations, c) provide the means to observe spatial data skills through personal interaction and to create shared mental models and experiences across organisational boundaries, and d) enable the combination of new and existing spatial data sets from different organisations to create new knowledge.

6) Finally, the results revealed that the government initiative called 'integrated development planning' has emerged as another incentive for spatial data

sharing. Since this has not previously been linked to the promotion of spatial data sharing, specific reference to the benefits of spatial data sharing in terms of integrated development planning should be made in the context of the spatial data infrastructure initiative in South Africa.

Theoretical Considerations

The literature on spatial data sharing includes an array of concepts and anecdotal evidence that aim to explain why people or institutions do or do not share spatial data. A behavioural structural model was developed to explore, organise and test some of the underlying facets of spatial data sharing behaviour (i.e., outcomes, social pressures and expectations, and resources and opportunities).

The quantitative analysis of the empirical data in Chapter 7 verified that the model for measuring the willingness of organisations to share spatial data developed in Chapter 5 is valid and reliable. The assessment of the validity of the model constituted a crucial step in the approach of this study which set out to systematise the determinants of organisations' spatial data sharing behaviour. Only when the model had been found to be valid for measuring the constructs under investigation, could it be relied upon to further the understanding of the incentives and disincentives experienced by key individuals within organisations to engage in spatial data sharing across organisational boundaries.

In this section, the contributions of the model to the understanding of the diffusion of the GIS technological innovation are outlined. The extent to which it offers a means of examining the social or behavioural aspects of the diffusion process more generally in other contexts of GIS usage and in studies of the diffusion of innovations in the science and technology policy research field are also discussed.

Diffusion research seeks to explain how and why an innovation spreads within a given group of adopters. The focus of this research has been on the preconditions for the diffusion of GIS such that the users of the technology will have a greater likelihood to behave in ways that may increase the potential benefits of the application of GIS. One precondition for the use of GIS is the availability and accessibility of appropriate spatial data as they are indispensable for the use of GIS. Spatial data sharing across organisational boundaries is conceived as a way of reducing duplication, inefficiencies, and lost opportunities for using spatial data in GIS. This study has provided a systematic and quantitative analysis of the conditions under which the different actors involved with spatial data may be willing or unwilling to share these data across organisational boundaries. The results of this approach have provided a basis for specific guidance as to how policy makers may influence the actual behaviour of spatial data sharing more effectively and, therefore, the model has contributed to a deeper understanding of the preconditions for the more effective use of GIS, an information and technology application that is being considered a useful tool for decision making in development planning.

The model developed in this research is relevant to other related contexts of GIS usage, such as other national contexts. As was shown in Chapter 2, spatial data infrastructure initiatives are emerging at the national level in the industrialised and developing countries to overcome obstacles to greater data availability. This study has shown that the model provides the basis for explaining the motivations for spatial data sharing behaviour within a specific context, in this case, South Africa. Based on the validation of the model, it can be argued that the model as implemented using the questionnaire instrument, could also be applied in other national contexts to elicit the dispositions of actors in the local GIS community and to provide a basis for specific guidance as to how policy makers might influence the actual behaviour of spatial data sharing more effectively.

At a broader level of application, the research contributes to studies of the diffusion of innovations in the science and technology policy research field. The behavioural aspects of the process of innovation have been examined by Montalvo Corral (2002). However, the process of diffusion of a new technological innovation may hinge on a number of preconditions that determine the extent to which the new technology is put to productive use (Sarkar, 1998). The perspective adopted in this research was to consider the social or behavioural nature of these preconditions. Systemic factors such as the availability of materials and components may play a role in the diffusion of an innovation (Freeman and Soete, 1997). Rogers has placed emphasis on the social aspects of the diffusion process, arguing that 'the diffusion of innovations is essentially a social process' (Rogers, 1995: xvii). The main elements in the diffusion of innovations, according to Rogers, are the characteristics of the innovation, communication channels, time and the members of a social system. By focusing on the decisions and behaviour of these members, the social psychological approach adopted here can be applied to contribute to a better understanding of the social or behavioural aspects that may affect the preconditions of the diffusion of technological innovations, such as the availability of inputs.

A Reflection on the Research

This section considers how this study departs from previous research by reflecting on the specific methods used to implement the theoretical approach. An alternative approach is also discussed.

Methods Employed

The research presented here differs from previous studies of spatial data sharing in a number of ways. First, a multi-disciplinary approach was used, with the organising framework drawn from social psychology. The use of an organising framework also meant that several bodies of research could be drawn upon to explore the phenomenon under investigation in a systematic fashion. Furthermore, the research employed a combination of qualitative and quantitative empirical analysis. Finally, the methodology was unique in that it used face-to-face interviews to implement the survey, complemented by an accidental sampling technique to produce a reliable representation of the population. This procedure also served as a screening mechanism to ensure that the 'right' person, i.e. the decision maker with respect to the spatial data sharing issue, in each organisation was completing the questionnaire.

In the application of the TPB, the sole reliance on belief elicitation using interviews to arrive at a model of a particular behaviour has been called into question with regard to the completeness of covering relevant beliefs (Conner and Armitage, 1998). The complementary approach taken in this research of relying on secondary material and different bodies of research, in addition to the beliefs elicited in Phase I, proved successful in composing the model and operationalising it in the questionnaire instrument, confirming the experience of others (e.g., Taylor and Todd, 1995). In this study, interviews alone would not have been sufficient to provide information about the full range of beliefs or to provide a basis for translating them into measures and questionnaire items.

Alternative Approach to Examining Spatial Data Sharing Behaviour

The Theory of Planned Behaviour is an extension of the Theory of Reasoned Action (Fishbein and Ajzen, 1975; Ajzen and Fishbein, 1980), the difference between the two being that the TPB includes the 'perceived control' construct to examine behaviour that cannot *a priori* be assumed to be under volitional control. One of the findings of this study was that the technical aspects of sharing in terms of skills and resources considered under 'perceived control' could not be confirmed as an important factor in influencing the willingness of individuals embedded in organisations to engage in spatial data sharing. It might therefore be argued that the TRA, which includes only the 'attitude' and the 'social pressure' constructs, would have sufficed to explain spatial data sharing behaviour. However, two objections can be raised to this argument.

First, the TPB is appropriate for explaining behaviour when it is not known in advance whether and to what extent the performance of the behaviour is reliant on particular skills, resources and opportunities. The starting point for this research was to argue that current spatial data sharing initiatives focus on the resolution of technical obstacles to spatial data sharing at the expense of taking social and behavioural aspects into consideration. Had this study been carried out applying the TRA instead of the TPB, it would not have been possible to establish empirically that the technical aspects that feature so prominently in the current spatial data sharing initiatives do not play such an important role in influencing the willingness of decision makers in organisations to engage in spatial data sharing across organisational boundaries. As it stands, the results of this research strongly support the conclusion that social and behavioural aspects also need to be considered if greater spatial data sharing is to be achieved.

Second, the model developed in this study can be readily applied in combination with a complete assessment of spatial data sharing behaviour *subsequent* to administering the questionnaire (as compared to the self-reports of past sharing behaviour employed in this research). In that instance, it would be possible to examine the nature of the direct link between 'perceived control' and sharing behaviour postulated by the TPB to be able to *predict* as well as *explain* spatial data sharing behaviour.

Directions for New Areas of Research

The issues addressed here would benefit from further research in two areas. First, the empirical research was carried out in a single country, South Africa. Hence, future research is required to determine whether the patterns of attitudes, beliefs, and behaviour that were revealed in this study for the case of South Africa are likely to vary substantially in other national contexts. Spatial data infrastructure initiatives are emerging at the national level in the industrialised and developing countries and it would be fruitful to apply the model developed in this research in other national contexts to provide a basis for specific guidance as to how policy makers can foster spatial data sharing behaviour more effectively within their national context.

A closely related, second area of new research arises from the drive towards greater co-ordination of spatial data infrastructures at regional and global levels. A comparative study of several countries, for example, in Europe within the context of INSPIRE (the INfrastructure for SPatial InfoRmation in Europe) or, more internationally still, within the context of the GSDI initiative, could be based on the model developed in this research. This would help to identify the differences in incentive structures that otherwise may not be addressed and, thus, assist further progress of such initiatives.

Conclusion and Final Remarks

While GIS can serve as useful tools to support decision making in a range of development activities, the potential benefits of their application continue to be constrained - particularly in developing countries - by the limited resources that are available for generating spatial data. If the key players involved in the use and supply of spatial data in different organisations are willing, spatial data sharing across organisational boundaries can provide a way of greatly reducing the costs of using GIS effectively.

The process of establishing a spatial data sharing 'culture' which is the central theme that has been addressed in this research is just one component of a more complex set of interactions between the users and suppliers of spatial data. These interactions also include trends in many areas that are occurring in the international policy environment. A particularly important trend that is likely to affect the potential for expansion of data sharing is the treatment of copyright on digital information, the enforcement of intellectual property rights legislation, and the way these developments influence the use and exchange of spatial data sets. Thus far, most research and policy discussion focuses on the more technical aspects of spatial data sharing. The need to consider the non-technical aspects contributing to resistance to spatial data sharing has been demonstrated in this research. The failure so far to address these aspects may be attributable to the perception on the part of the stakeholders that the technical issues are more easily resolved than those related to social and behavioural aspects of data sharing. This study has provided a methodology and evidence that may be expected to provide a foundation for further consideration of these very important aspects of the use of GIS as a tool for development.

Phase I Interview Questions

This appendix to Chapter 4 contains the questions for the semi-structured interviews carried out in Phase I of the empirical research to elicit the beliefs.

Behavioural Beliefs

1. What do you see as the advantages/ gains/ benefits of your organisation's engagement in spatial data sharing?
2. What do you see as the disadvantages/ drawbacks of your organisation's engagement in spatial data sharing?
3. Is there anything else that you associate with spatial data activities of your organisation?

Normative Beliefs

1. Are there any people or institutions who you think want your organisation to /push your organisation to engage in spatial data sharing with other organisations?
2. Are there any people or institutions who you think *don't* want your organisation to engage in spatial data sharing across organisational boundaries?
3. Does anybody else come to mind when you think about your organisation participating in spatial data sharing across organisational boundaries?

Control Beliefs

1. What kind of skills or abilities do you think your organisation needs to engage in spatial data sharing across organisational boundaries?
2. What does your organisation need to know to engage in spatial data sharing across organisational boundaries?
3. What experience do you think your organisation needs to engage in spatial data sharing across organisational boundaries?
4. What kind of information (non-spatial) do you think your organisation needs to engage in spatial data sharing across organisational boundaries?

5. What kind of technological facilities do you think your organisation needs for *sharing* spatial data with other organisations rather than just *using* spatial data?

6. What additional resources in terms of time/ money do you think your organisation needs for *sharing* spatial data as opposed to just *using* spatial data?

7. Are there any people or institutions who your organisation needs to help it/ who are important/ who are facilitators to take part in spatial data sharing activities with other organisations?

8. Are there any institutions or people who you think are stopping or preventing your organisation from carrying out those sharing activities?

9. Are there any particular circumstances/ opportunities you think your organisation relies on for sharing spatial data across organisational boundaries?

10. Are there any hurdles or constraints you think are stopping your organisation from carrying out those sharing activities?

Questionnaire Items

This appendix contains the items of the questionnaire instrument. The format of the scales used in the actual questionnaire instrument is indicated in Chapter 5.

Intention Items

1. In our organisation, we definitely intend to get involved in spatial data sharing. (agree – disagree)
2. Our organisation intends to engage in spatial data sharing. (agree – disagree)
3. Considering the outcomes, pressures, capabilities and resources, the likelihood for your organisation to engage in spatial data sharing is ...: (low- high)

Direct Measures

Attitude

1. The consequences of sharing spatial data for our organisation's resources, qualitative work aspects, and strategic position, and in terms of social benefits would be ...: (positive – negative)
2. The benefits resulting from sharing spatial data for our organisation would be ...: (small – large)
3. In our organisation, we consider spatial data sharing to be ...: (please mark all the scales) (threatening- opportune) (bad – good) (negative – positive) (detracting – rewarding)

Social Pressure

1. People and institutions that are important to the success of our organisation think we should engage in spatial data sharing. (agree – disagree)
2. The social pressures (organisational, GIS community, market, institutions and social beneficiaries) received by this organisation to engage in spatial data sharing activities are likely to be ...: (weak – strong)

Perceived Control

1. After considering our organisation's technological and interpersonal skills, resources, influence and dependence on sharing partners, and the opportunities created by the NSIF, it seems that spatial data sharing is likely to be ...: (easy – difficult)
2. It is mainly up to us whether our organisation engages in spatial data sharing activities. (agree – disagree)

Domain Level Items

Attitude

1. In terms of our organisation's resources (time, money, labour, data storage, quality and quantity of spatial data), spatial data sharing would result in ...: (costs – benefits)
2. The consequences for the qualitative aspects of our organisation's work resulting from our spatial data sharing activities are likely to be ...: (positive – negative)
3. In terms of our organisation's strategic position, spatial data sharing would result in ...: (gains – losses)
4. The social benefits of our organisation's engagement in spatial data sharing would be ...: (small – large)

Social Pressure

1. Overall, the pressure of the GIS community regarding our organisation's sharing activities is likely to be ...: (weak – strong)
2. Overall, the pressures our organisation is receiving from the spatial data/ GIS market regarding our sharing activities are likely to be ...: (weak – strong)
3. Overall, the institutional pressures (NSIF, spatial data agreements, politicians) on our organisation regarding spatial data sharing activities are likely to be ...: (weak – strong)
4. Overall, the organisational pressures on our organisation (from other departments in our organisation, our organisational goals/mission, management, sharing 'champions') regarding our spatial data sharing activities are likely to be ...: (weak – strong)
5. Overall, the social obligations for our organisation to engage in spatial data sharing activities are likely to be ...: (weak – strong)

Perceived Control

1. For spatial data sharing, our organisation is likely to have the necessary technical capabilities (in spatial data, metadata, and computing). (agree – disagree)
2. For spatial data sharing, our organisation is likely to have the necessary experience and interpersonal skills in networking, teamworking, finding appropriate (willing, reliable, compatible) sharing partners and negotiating sharing arrangements (win-win, price, ownership, liability). (agree – disagree)
3. In our organisation, we are likely to have the necessary resources (human and financial resources and clear guidelines) for spatial data sharing. (agree – disagree)
4. In our organisation, the extent to which we would be able to exert influence over spatial data sharing in terms of the dependence on, and control over, spatial data would be ...: (small – large)
5. For our organisation, the extent to which we (would) rely on the NSIF activities to create opportunities for spatial data sharing is likely to be ...: (small – large)

Attitude Items

Please consider the following resource-related cost implications of spatial data sharing. (insignificant costs – significant costs)
1. Time required to locate spatial data ex-house
2. Trained staff required to exchange and integrate spatial data
3. Data storage required for spatial data to be shared
4. Administrative efforts for sharing (e.g. for strategic policy development, contracts, agreements, copyright negotiations)
5. Expenses incurred from spatial data sharing (time, equipment, labour, data cost)

Please consider the following resource-related benefits of spatial data sharing. (insignificant savings – significant savings)
1. Time saved otherwise needed to capture spatial data in-house
2. Trained staff saved otherwise needed for capturing and maintaining spatial data
3. Data storage saved otherwise needed for divergent/ duplicated databases
4. Administrative efforts saved otherwise needed for data capture
5. Expenses saved otherwise needed for data capture (time, equipment, data storage, labour)

How likely is it that your organisation's engagement in spatial data sharing would result in the following resource-related cost implications? (likely – unlikely)
1. Sharing would require time to locate spatial data *ex-house*

2. Sharing would require trained staff for spatial data exchange / integration
3. Sharing would require data storage for spatial data to be shared
4. Sharing would require administrative efforts
5. Sharing would result in expenses (time, equipment, labour, data cost)

How likely is it that your organisation's engagement in spatial data sharing would have the following resource-related benefits? (likely – unlikely)
1. Sharing would save time otherwise needed to capture spatial data in-house
2. Sharing would save trained staff otherwise needed for capturing and maintaining spatial data
3. Sharing would save data storage otherwise needed for divergent/ duplicated databases
4. Sharing would save administrative efforts otherwise needed for data capture

5. Sharing would save expenses otherwise incurred for data capture

The following outcomes for spatial data would be ...: (positive – negative)
1. Increased quantity of spatial data available to your organisation
2. Introduction of standards agreed with other organisations (spatial and attribute data standards, metadata standards)

How likely is it that your organisation's engagement in spatial data sharing would result in the following outcomes? (likely – unlikely)
1. Sharing would increase the quantity of spatial data available to your organisation
2. Sharing would result in the introduction of standards agreed with other organisations (spatial and attribute data standards, metadata standards)
3. Sharing would result in quality improvements of shared spatial data sets (quality of positional accuracy, attribute accuracy, consistency, completeness, and lineage)
4. Sharing would result in the introduction of errors and gaps in shared spatial data sets

The way in which our organisation's spatial data sharing activities would impact the following qualitative work aspects would be ...: (positive – negative)
1. Focus on core activities
2. Usefulness of our GIS
3. Quality of decision-making

Please consider the following aspects of losing control over spatial data for your organisation's strategic position. (threat – opportunity)
1. For our organisation's strategic position, not copyrighting the spatial data sets our organisation (would) make available to other organisations would present a...:

2. For our organisation's strategic position, unregulated access to our organisation's spatial data by other organisations would present a ...:
3. For our organisation's strategic position, use of our organisation's spatial data by organisations other than our sharing partners would present a ...:
4. For our organisation's strategic position, not enforcing any rules regarding the ownership of, access to and use of, our organisation's spatial data would present a ...:

In terms of your organisation's strategic position, please consider the likelihood of the following aspects of losing control over your organisation's spatial data as a result of engaging in spatial data sharing. (likely – unlikely)
1. How likely is it that spatial data sets your organisation makes available to other organisations would not be covered by copyright?
2. How likely is it that access to your organisation's spatial data by other organisations would not be regulated in some way?
3. How likely is it that organisations other than your sharing partners would be able to use spatial data owned by your organisation?
4. How likely is it that rules regarding the ownership of, access to and use of, your organisation's spatial data could not be enforced?

Spatial data sharing can create threats and opportunities for the strategic position of all parties involved - the strategic position of your organisation and that of the organisations you may be sharing with. Please indicate how you value the following consequences of sharing spatial data for your organisation's strategic position. (threat – opportunity)
1. If sharing were to enable the combination of new and existing spatial data sets from different organisations to create new knowledge, in terms of our organisation's strategic position this would present a ...:
2. If sharing were to allow access to ideas that are codified in spatial data (e.g. as thematic layers) and their translation, combination and application in different organisations, in terms of our organisation's strategic position this would present a ...:
3. If sharing were to provide the means to observe spatial data skills in personal interaction and to create shared mental models and experiences across organisational boundaries, in terms of our organisation's strategic position this would present a ...:
4. If sharing were to trigger dialogue and collective reflection across organisational boundaries (e.g. joint problem solving) and to result in articulating, making explicit and comparing spatial data-related concepts, in terms of our organisation's strategic position this would present a ...:
5. If sharing were to increase interdependence among organisations, in terms of our organisation's strategic position this would present a ...:
6. If sharing were to redistribute influence among organisations, in terms of our organisation's strategic position this would present a ...:

How likely is it that spatial data sharing would have the following consequences for the strategic position of your organisation? (likely – unlikely)
1. Sharing would enable the combination of new and existing spatial data sets from different organisations to create new knowledge.
2. Sharing would allow access to ideas that are codified in spatial data (e.g. as thematic layers) and their translation, combination and application in different organisations.
3. Sharing would provide the means to observe spatial data skills in personal interaction and to create shared mental models and experiences across organisational boundaries.
4. Sharing would trigger dialogue and collective reflection across organisational boundaries and result in articulating, making explicit and comparing spatial data-related concepts. (e.g. joint problem solving)
5. Sharing would increase interdependence among organisations.
6. Sharing would redistribute influence among organisations.

Our organisation's spatial data sharing activities would result in the following social benefits ...: (likely – unlikely)
1. Integrated development planning
2. Greater benefits to society at large

Social Pressure Items

The following people think that our organisation's engagement in spatial data sharing would be ...: (positive – negative)
1. GIS departments in local authorities
2. GIS departments in provincial government
3. GIS departments in national government
4. GIS departments in para-statal organisations
5. GIS users in the private sector
6. NGO GIS users
7. GIS users in academic research institutions

In general, our organisation very much wants to do what the following people want us to do ...: (not at all – very much)
1. GIS departments in local authorities
2. GIS departments in provincial government
3. GIS departments in national government
4. GIS departments in para-statal organisations
5. GIS users in the private sector
6. NGO GIS users
7. GIS users in academic research institutions

The following people think that our organisation's engagement in spatial data sharing would be ...: (positive – negative)
1. Commercial spatial data brokers
2. Public spatial data providers
3. Private spatial data providers
4. GIS suppliers

In general, how much does your organisation want to do what the following people want you to do? (not at all – very much)
1. Commercial spatial data brokers
2. Public spatial data providers
3. Private spatial data providers
4. GIS suppliers

The following institutions consider that our organisation's engagement in spatial data sharing would be ...: (positive – negative)
1. National Spatial Information Framework (NSIF)
2. Spatial data agreement(s) applicable to our organisation
3. Politicians

In general, how much does your organisation want to comply with the expectations of the following institutions? (not at all – very much)
1. National Spatial Information Framework (NSIF)
2. Spatial data agreement(s) applicable to our organisation
3. Politicians

The following (groups of) people think that our organisation's engagement in spatial data sharing would be ...: (positive – negative)   .
1. Other departments in our organisation
2. Our management
3. Certain 'champions' for spatial data sharing in our organisation

In general, how much does your organisation want to do what the following people want you to do? (not at all – very much)
1. Other departments in our organisation
2. Our management
3. Certain 'champions' for spatial data sharing in our organisation

Please indicate what your organisational goal / mission mandates with regard to your organisation's engagement in spatial data sharing activities. (should share; should not share)

In general, to what extent do the people in your organisation want to act in accordance with your organisational goals / mission? (not at all – very much)

The following social beneficiaries suggest that our organisation's engagement in spatial data sharing would be desirable. (agree – disagree)
1. Integrated development planning
2. Society at large

How willing is your organisation to engage in spatial data sharing just for the sake of the following beneficiaries? (not at all – extremely)
1. Integrated development planning
2. Society at large

Perceived Control Items

How important are the following spatial data-related skills for sharing spatial data? (important – not at all)
1. Assessing the quality of spatial data (positional accuracy, attribute accuracy, consistency, completeness, and lineage)
2. Handling different formats of spatial data
3. Mastering different standards (for spatial and attribute data)
4. Selecting spatial data from a database (e.g. as single observations, subsets, or a thematic layer)
5. Integrating spatial data from diverse sources (incl. converting and manipulating data, verifying its geometric and attribute quality, maintaining its consistency with related data, and updating layers independently of each other)

For your organisation, how difficult are the following spatial data-related activities? (easy – difficult)
1. Assessing the quality of spatial data
2. Handling different formats of spatial data
3. Mastering different standards (for spatial and attribute data)
4. Selecting spatial data from a database
5. Integrating spatial data from diverse sources

How important are the following mtadata-related skills for sharing spatial data? (important – not at all)
1. Interpreting metadata
2. Using metadata interfaces / catalogues
3. Capturing metadata
4. Applying metadata standards
5. Maintaining and updating metadata

For your organisation, how difficult are the following metadata-related activities? (easy – difficult)
1. Interpreting metadata
2. Using metadata interfaces / catalogues
3. Capturing metadata

4. Applying metadata standards
5. Maintaining and updating metadata

How important are the following general computing skills for sharing spatial data? (important – not at all)
1. Database administration skills
2. Using the Internet to locate spatial data sources
3. Using the Internet to distribute spatial data
4. Transferring spatial data to and from different media (e.g. CD-Rom)

How difficult are the following computing tasks for your organisation? (easy – difficult)
1. Database administration
2. Using the Internet to locate spatial data sources
3. Using the Internet to distribute spatial data
4. Transferring spatial data to and from different media

How important are the following interpersonal skills for sharing spatial data? (important – not at all)
1. Establishing and fostering a network of contacts
2. Identifying and attending meetings to network with people
3. Keeping a finger on the pulse of a network in terms of what others are working on and letting them know about your own work
4. Collaborating with others
5. *Interdisciplinary teamworking*

In our organisation, dealing with the following interpersonal aspects is ...: (easy – difficult)
1. Establishing and fostering a network of contacts
2. Identifying and attending meetings to network with people
3. Keeping a finger on the pulse of a network
4. Collaborating with others
5. Interdisciplinary teamworking

How important are the following characteristics of any current or prospective spatial data sharing partner? (important – not at all)
1. Willingness, responsiveness
2. Reliability
3. Compatible purpose of application of the spatial data (use of data within specific context)
4. 'Organisational fit' (type of organisation you are comfortable with)

For our organisation, finding sharing partners with the following characteristics would be ...: (easy – difficult)
1. Willing, responsive
2. Reliable

3. Compatible purpose of application
4. Compatible 'organisational fit'

For spatial data sharing, how important is it to negotiate the following characteristics of any spatial data sharing arrangement? (important – not at all)
1. 'Win-win' situation (arrangement that equally benefits all parties involved)
2. Pricing of spatial data (e.g. licence fee, marginal cost of reproduction, barter, free of charge)
3. Data ownership agreements such as copyright
4. Liability agreement (that mandates some acceptable level of performance in the delivery of services or the quality of products)

For our organisation, negotiating the following terms and conditions of any sharing arrangement would be ...: (easy – difficult)
1. 'Win-win' situation
2. Price of spatial data
3. Data ownership agreements such as copyright
4. Liability agreement

Please indicate the kind of experience your organisation has with spatial data sharing. (agree- disagree)
1. In the past, our organisation has been heavily involved in spatial data sharing with other organisations.
2. In the past, our organisation has had bad experiences with spatial data sharing with other organisations.
3. We do not have a clear overview of the consequences for our organisation of spatial data sharing with other organisations.

How important are the following resources for spatial data sharing?
1. Sufficient number of staff for spatial data sharing activities
2. Time to spend on spatial data sharing activities
3. Availability of funding
4. Organisational guidelines related to spatial data (e.g. sharing-related policy)

In our organisation, the extent to which we have control over the following resources is ...: (small – large)
1. Sufficient number of staff for spatial data sharing activities
2. Time to spend on spatial data sharing activities
3. Availability of funding
4. Organisational guidelines related to spatial data

Please evaluate the following dependence aspects of sharing in terms of the importance of spatial data. (important – not at all)
1. How important is the spatial data your organisation (would) receive from sharing partners for the continued operation of your organisation?

2. How important is the spatial data your organisation (would) give to sharing partners for their continued operation?

Please evaluate the following dependence aspects for spatial data made accessible to your organisation. (important – not at all)
1. How important is it that the spatial data sets your organisation receives from other organisations **not** be covered by copyright?
2. How important is it that access to the spatial data your organisation requires from other organisations **not** be regulated in some way other than ownership? *(e.g. formal access policies, licence restriction)*
3. How important is it that your organisation is able to use spatial data that you do **not** own?
4. How important is it that rules regarding the copyright of, access to and use of, spatial data from other organisations **not** be enforced?

Please consider the likelihood of the following dependence aspects for spatial data made accessible to your organisation. (small – large)
1. The extent to which other organisations would be able to ensure that the spatial data sets your organisation receives from them are covered by copyright is likely to be ...:
2. The extent to which other organisations would be able to ensure that access to the spatial data your organisation requires from them is regulated in some way is likely to be ...:
3. The extent to which other organisations would be able to ensure that your organisation does not use spatial data owned by them is likely to be ...:
4. The extent to which other organisations would be able to enforce rules regarding the copyright of, access to and use of, their spatial data by your organisation is likely to be ...:

Please evaluate the following control aspects for spatial data your organisation would make accessible to other parties. (important – not at all)
1. How important is it that the spatial data your organisation (would) make available to other organisations be covered by copyright?
2. How important is it that access to your organisation's spatial data by other organisations be regulated in some way other than ownership?
3. How important is it that spatial data you organisation owns cannot be used by organisations other than your sharing partners?
4. How important is it that rules be enforced regarding the copyright of, access to and use of, your organisation's spatial data?

Please consider the likelihood of the following control aspects for spatial data your organisation would make accessible to other parties. (small – large)
1. The extent to which your organisation could cover the spatial data it would make available to other organisations by copyright is likely to be ...:
2. The extent to which your organisation could regulate access to its spatial data by other organisations in some way other than ownership is likely to be ...:

3. The extent to which your organisation could ensure that spatial data owned by it cannot be used by organisations other than your sharing partners is likely to be ...:
4. The extent to which your organisation is able to enforce rules regarding the copyright of, access to and use of, its spatial data is likely to be ...:

Please evaluate the following dependency aspects in terms of spatial data supplies. (agree – disagree)

1. Our organisation has many alternative sources for the spatial data we require.
2. The alternative sources for the spatial data that our organisation requires are likely to be extremely stable.
3. Any organisation our organisation (would) make its spatial data accessible to has many alternative sources for this data.
4. The alternative sources for the spatial data our organisation (would) make accessible to other organisations are likely to be extremely stable.

How self-sufficient is your organisation in terms of spatial data resources? (dependent – self-sufficient)

At the moment, the position of our organisation in terms of spatial data exchanges can be considered as ...: (giving – receiving spatial data)

How important would be the following activities by the National Spatial Information Framework (NSIF) for your organisation's spatial data sharing activities? (important – not at all)

1. NSIF awareness creation
2. Fora organised by the NSIF
3. Policy development
4. Standards alignment
5. South African core data set identification
6. Spatial Data Discovery Facility (searchable on-line clearinghouse with Internet links to distributed sites)

Past Spatial Data Sharing Items

Type of your Organisation

☐   private sector / commercial
☐   academic institution
☐   non-governmental organisation (NGO)
☐   local authority
☐   provincial government
☐   national government
☐   para-statal organisation
☐   other, please specify: _____

Spatial data role of your organisation (please mark those that apply)

- ☐ spatial data recipient (end user)
- ☐ spatial data supplier
- ☐ spatial data broker
- ☐ other, please specify: _____

Type(s) of spatial data exchange(s) of your organisation (please mark those that apply)

- ☐ data supplier ↔ end user
- ☐ data supplier ↔ data broker
- ☐ end user ↔ data broker
- ☐ between data suppliers
- ☐ between end users
- ☐ between data brokers
- ☐ other, please specify: _____
- ☐ not applicable

Type(s) of spatial data sharing partner(s) of your organisation (please mark those that apply)

- ☐ vendor
- ☐ private sector / commercial
- ☐ academic institution
- ☐ non-governmental organisation (NGO)
- ☐ local authority
- ☐ provincial government
- ☐ national government
- ☐ para-statal organisation
- ☐ other, please specify: _____
- ☐ not applicable

Schedule of spatial data exchange(s)

- ☐ on schedule (regularly)
- ☐ on demand (ad hoc)
- ☐ other, please specify: _____
- ☐ not applicable

Frequency of spatial data exchange(s)

- ☐ daily
- ☐ weekly
- ☐ monthly
- ☐ yearly
- ☐ project basis
- ☐ once-off
- ☐ never

Most recent spatial data exchange(s)

- ☐ this week
- ☐ last week
- ☐ last month
- ☐ last 6 month
- ☐ more than 6 months ago
- ☐ not applicable

Sharing arrangement(s) (please mark those that apply)

- ☐ informal / voluntary
- ☐ formal contract
- ☐ mandate to share or distribute
- ☐ profit-making venture
- ☐ not applicable

Charges for shared spatial data (please mark those that apply)

- ☐ free of charge
- ☐ barter
- ☐ license fee
- ☐ consortium membership
- ☐ marginal cost of reproduction
- ☐ cost recovery basis
- ☐ market value
- ☐ not applicable

Average quantity

- ☐ single observation
- ☐ summary / aggregate observations
- ☐ selected subset
- ☐ theme
- ☐ entire database
- ☐ not applicable

Analysis of Differences in Intentions

This appendix to Chapter 6 presents the results of the analysis of differences in intentions. In order to compare the intentions for different groups, for example, the different sectors in the GIS community, the Levene test was carried out first to find out if it is appropriate to use analysis of variance (one-way ANOVA) to test for mean differences among the groups. The analysis of variance is based on the assumption that the variances of the groups do not differ too widely. When this assumption was violated (indicated by a significant Levene test result), the Kruskal-Wallis test was used to compare the groups; otherwise a one-way ANOVA was carried out. If the result of the Kruskal-Wallis or the one-way ANOVA (the ANOVA F) is significant, there are significant differences in intentions.

Analysis of Differences in Intention between Sectors in the GIS Cmmunity

The result of the Levene test is displayed in Table 1.

Table 1   Levene test results (GIS community sectors)

Test of Homogeneity of Variances			
INT3.1			
Levene Statistic	df1	df2	Sig.
4.160	8	103	.000

This shows that the Levene test of homogeneity of variances among groups was significant (p=.000) so that the assumption of equal variances was violated. This means that the variances for the sectors do differ significantly. Hence, a Kruskal-Wallis test had to be carried out (see Table 2).

Table 2   Kruskal-Wallis results (GIS community sectors)

Test Statistics[a,b]

	INT3.1
Chi-Square	10.793
df	8
Asymp. Sig.	.214

a. Kruskal Wallis Test

b. Grouping Variable: sector

The result of the Kruskal-Wallis test shows that there was no overall difference between the intentions of the respondents representing the different sectors (p=.214).

Analysis of Differences in Intention According to Reliance on External Spatial Data

The respondents were divided into three groups according to their response regarding their organisation's reliance on external spatial data (dependent, self-sufficient, neutral). The Levene test of homogeneity of variances among groups (shown in Table 3) was not significant (p=.681) so that the assumption of equal variances was not violated and the one-way ANOVA to compare the intention means of the groups could be used. The results of this procedure are shown in Table 4.

Table 3   Levene test results (reliance on external spatial data)

Test of Homogeneity of Variances

INT3.1

Levene Statistic	df1	df2	Sig.
.385	2	109	.681

Table 4   One-way ANOVA test results (reliance on external spatial data)

ANOVA					
INT3.1					
	Sum of Squares	df	Mean Square	F	Sig.
Between Groups	.319	2	.159	.071	.931
Within Groups	244.785	109	2.246		
Total	245.104	111			

Since the ANOVA F test was not significant (p=.931), it can be concluded that there was no overall difference between the intentions of the respondents according to their organisation's reliance on external spatial data.

Analysis of Differences in Intention According to Spatial Data Position

The respondents were divided into three groups according to their response regarding their organisation's spatial data position in spatial data exchanges (receiving more spatial data, giving more spatial data, neutral). The Levene test of homogeneity of variances among groups (shown in Table 5) was significant (p=.050) so that the assumption of equal variances was violated and the Kruskal-Wallis test to compare the intention means of the groups had to be used. The results of this procedure are shown in Table 6.

Table 5   Levene test results (spatial data position)

Test of Homogeneity of Variances			
INT3.1			
Levene Statistic	df1	df2	Sig.
3.084	2	109	.050

Table 6   Kruskal-Wallis results (spatial data position)

Test Statistics[a,b]

	INT3.1
Chi-Square	1.416
df	2
Asymp. Sig.	.493

a. Kruskal Wallis Test

b. Grouping Variable: SDPOS

As the Kruskal-Wallis test was not significant (p=.493), it can be concluded that there was no overall difference between the intentions of the respondents according to their organisation's spatial data position in spatial data exchanges.

Descriptive Statistics of Actual Sharing Behaviour

This appendix to Chapter 6 contains additional descriptive statistics for the self-reports of spatial data sharing behaviour. Specifically, it provides details about who is sharing with whom and the terms and conditions under which spatial data is shared.

Who is Sharing with Whom

Figure 1 shows the role of the organisations in the sample (as reported by the respondents) with respect to their involvement with spatial data. In all of the sectors, apart from the NGOs, there are spatial data recipients, suppliers and brokers.

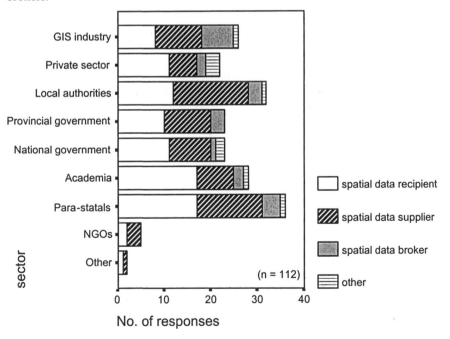

Figure 1   **Spatial data role of the organisation by sector**

Figure 2 indicates the types of spatial data exchanges that the organisations engage in. The most frequent exchanges were indicated between data suppliers and end users, and among the end users themselves.

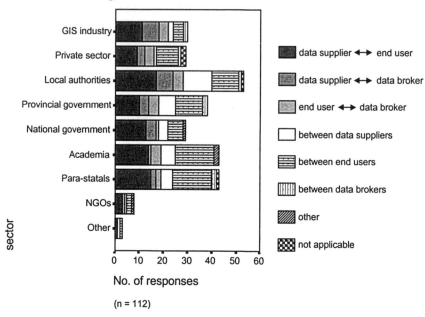

No. of responses

(n = 112)

Figure 2   Types of spatial data exchanges by sector

Figure 3 shows the types of sharing partners of the different GIS community sectors in the sample. At this aggregate level of the sectors (as opposed to the individual organisational level), it seems that everybody is sharing with everybody else.

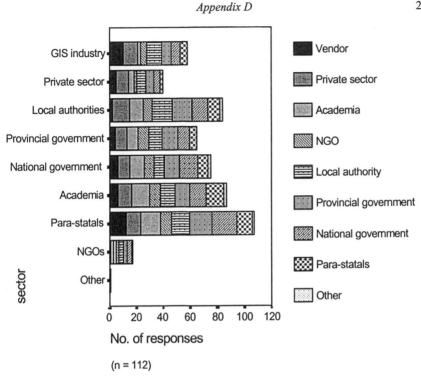

Figure 3   Types of sharing partners by sector

Terms and Conditions under which Spatial Data is Shared

Both formal and informal sharing arrangements were reported frequently by the respondents (see Figure 4). Only a quarter of the respondents indicated that their organisation has an explicit mandate to share.

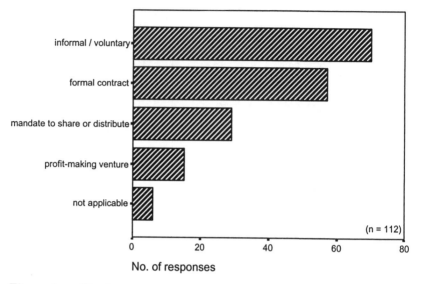

No. of responses

Figure 4   Sharing arrangements

Figure 5 gives an overview of the types of sharing arrangement that the different GIS community sectors in the sample engage in. Formal and informal arrangements were reported right across all of the sectors. As can be expected, profit-making ventures were reported by the GIS industry, para-statal organisations and national government (since the Surveyor General and other data producing organisations are included in this category) but they were also being used by some local authorities and academic institutions. Sharing arrangements that are based on a mandate to share were reported by all sectors but most frequently by the GIS industry, local and national government and para-statal organisations.

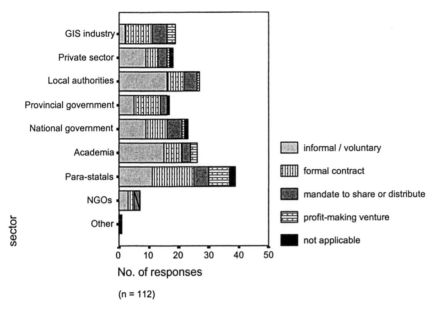

Figure 5   Types of sharing arrangements by sector

The most frequent response with respect to the charges for shared spatial data (see Figure 6), whether received or supplied by the organisations, was that spatial data was shared free of charge. Other frequent terms and conditions for shared spatial data entailed full cost recovery and marginal cost of reproduction. Full cost recovery refers to making spatial data available at a price that allows the originator to recover all the costs incurred in producing the data set. Prices set at the marginal cost of reproduction entails charging only for time and the medium (e.g. CD-Rom) required for providing a copy of the spatial data (sub)set.

Figure 7 presents the charges for shared spatial data by the different GIS community sectors in the sample. This shows that the different types of charges are used right across all of the sectors.

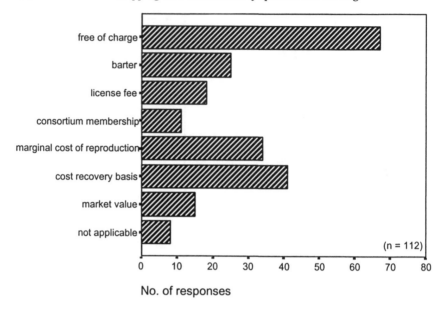

Figure 6   Charges for shared spatial data

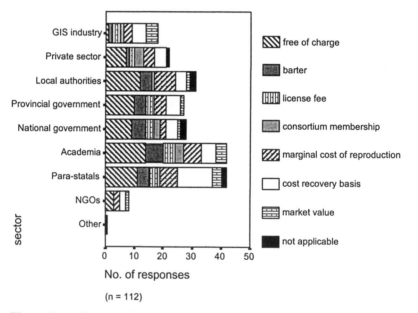

Figure 7   Charges for shared spatial data by sector

Appendix E

Empirical Validation of the Model

This appendix to Chapter 7 deals with the empirical validation of the willingness to share spatial data. Bryman (1989) has exposed the failure of most organisational research, and social science research more generally, to conduct and to report validity testing of the measures used. The aspects that need to be considered are the reliability and validity of the measure.

Reliability of measurement refers to *'the extent to which an instrument provides consistent measures'* (Aronson *et al.*, 1990: 350). Confirmation is needed about whether each scale of items is measuring a single idea (Bryman and Cramer, 1999). For the model of the willingness to share spatial data, this confirmation is needed at the level of the beliefs since these are grouped into domains and into sub-areas within some of the domains.

Two aspects of validity were considered important because the survey instrument was based on the model of the willingness to share spatial data across organisational boundaries using the TPB as an organising framework (see Chapter 5). First, *content validity* refers to the degree to which items tap the underlying, abstract concepts they are supposed to measure (Ghiselli *et al.*, 1981). The empirical data should therefore correspond to the theoretical basis of the model where different question items jointly are intended to measure the concepts suggested by the theory. The model of the willingness to share distinguishes between three main concepts: 'attitude', 'social pressure' and 'perceived control' (see Figure 1). The analysis of the model is required to assess the extent to which the different domains and the behavioural, normative and control beliefs seem to have been measuring each of the main concepts that they were intended to measure.

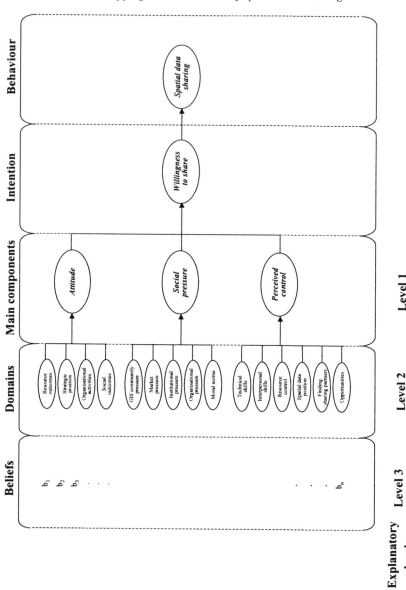

Figure 1   Hierarchical levels of the model

Second, *construct validation* consists of deducing and testing hypotheses from a relevant theory (Bryman and Cramer, 1999). Construct validation is necessary to confirm (or disconfirm) whether the theorised relationships that were postulated to hold between the measured variables in the model actually do hold; first, in terms of the relationship between the direct measures of the main components and the sums of the underlying beliefs and, second, in terms of the explanatory power of the model. The specific hypotheses developed in Chapter 4 were tested against the empirical data.

Reliability: Consistency of Measures

The first step in the validation of the model of spatial data sharing was to test the reliability of the measures employed. The most commonly used index of internal reliability is Cronbach's coefficient alpha (Bryman, 1989; Morgan and Griego, 1998; Cramer, 1998) to determine whether a scale that is made up of multiple items measures one particular characteristic. Cronbach's alpha is based on the average correlation of each item in the scale with every other item, thus indicating the extent to which items are related to each other. For a multiple item scale, the alpha coefficient should be above .70, whereas for scales with only a few items, a lower alpha is acceptable (Morgan and Griego, 1998).[46]

Applied to the items in the questionnaire, the alpha coefficient was calculated to assess the reliability of the items designed to measure the beliefs, grouped into domains and into areas within some of the domains, as shown in Figure 5.6 in Chapter 5. The results that were obtained for the groups of behavioural beliefs and outcome evaluations underlying 'attitude' are presented in Table 1. The results of the assessment of the belief measures for 'social pressure' and 'perceived control', are presented in Tables 2 and 3.

---

[46] High reliability of a scale does not guarantee that it is measuring what it is supposed to measure. It is necessary but not sufficient for validation (Aronson *et al.*, 1990). The extent to which items are measuring particular underlying concepts is assessed in the conceptual validation of the model using factor analysis.

**Table 1	Reliability of behavioural belief measures and outcome evaluations**

Construct (n = 112)	Reliability (Cronbach's alpha coefficient)	No. of items
Resource outcomes		
Magnitude of cost implications	.7491	5
Likelihood of cost implications	.7976	5
Magnitude of savings implications	.8478	5
Likelihood of savings implications	.8972	5
Outcomes for spatial data	.4460	2
Likelihood of outcomes for spatial data	.5475	4
Organisational activities		
Organisational activity outcomes	.7733	3
Strategic position outcomes		
Loss of control of spatial data	.9051	4
Likelihood of loss of control of spatial data	.7647	4
Knowledge creation outcomes	.8205	4
Likelihood of knowledge creation outcomes	.8088	4
Inter-organisational relations outcomes	.7144	2
Likelihood of inter-organisational relations outcomes	.7971	2
Social outcomes		
Likelihood of social outcomes	.8690	2

**Table 2	Reliability of belief-based social pressure measures**

Construct (n = 112)	Reliability (Cronbach's alpha coefficient)	No. of items in the questionnaire
GIS community pressure		
GIS community expectations	.8888	9
Compliance with GIS community referents	.9299	9
GIS community referents engagement in sharing	.8549	9
Market pressure		
Market expectations	.8590	4
Compliance with market referents	.8861	4
Institutional pressure		
Institutional expectations	.6037	3
Compliance with institutions	.7700	3
Organisational pressure		
Organisational expectations	.7550	3
Compliance with organisational referents	.6571	4
Moral Norms		
Moral norm pressure	.7057	2
Compliance with moral norms	.8805	2

Table 3   Reliability of belief-based perceived control measures

Construct (n = 112)	Reliability (Cronbach's alpha coefficient)	No. of items in the questionnaire
Technical skills		
Importance of spatial data skills	.6492	5
Availability of spatial data skills	.8367	5
Importance of Metadata skills	.9321	5
Availability of Metadata skills	.8733	5
Importance of general IT skills	.6689	4
Availability of IT skills	.7345	4
Interpersonal skills		
Importance of networking and collaboration skills	.8391	5
Availability of networking and collaboration skills	.8655	5
Importance of negotiation skills	.6648	4
Availability of negotiation skills	.7951	4
Past experience with sharing	.2611	3
Resource control		
Importance of resources for sharing	.6427	4
Control of over resources	.7806	4
Spatial data position		
Importance of spatial data	.7932	2
Importance of dependence aspects	.8070	4
Likelihood of dependence aspects	.8285	4
Importance of control over spatial data	.8527	4
Likelihood of control over spatial data	.8136	4
Availability and stability of spatial data supplies	.7341	4
Spatial data self-sufficiency	.5617	2
Finding sharing partners		
Importance of sharing partner characteristics	.5666	4
Availability of sharing partners	.8552	4
Opportunities		
Importance of opportunity creating activities	.9344	6

Similarly, for the 'direct' measures of intention, 'attitude', 'social pressure' and 'perceived control', inter-item reliability was assessed and these results are presented in Table 4.

Table 4   Reliability of 'direct' measures

Construct (n = 112)	Reliability (correlation coefficient*)	No. of items in the questionnaire
Intention (present)	*.7805*	3
Intention (near future)	*.8272*	3
Attitude (present)	*.8818*	6
Attitude (near future)	.5693	2
Social pressure (present)	.6585	2
Social pressure (near future)	.7441	2
Perceived control (present)	.1730	2
Perceived control (near future)	.1751	2

* In this list, reliability of the measures that are made up of more than 2 items is indicated by Cronbach's alpha coefficient (presented in *italics*).

The majority of the alpha coefficients for the scales in the questionnaire surpass the .70 criterion (as presented in Table 1-4).[47] It can be concluded that the measures in the questionnaire had highly satisfactory levels of internal reliability.

This suggests that the research procedure of eliciting beliefs and grouping them into domains and into sub-areas within some of the domains is appropriate for producing consistent scales. Furthermore, substantiating the consistency of the scales constitutes a necessary, albeit not sufficient, step in the validation of the model to establish a high degree of confidence in the results of the subsequent analysis of the determinants of the willingness to share spatial data. The further steps that are required are conceptual and construct validation of the model of spatial data sharing.

Conceptual Validation of the Model

The second stage of validation of the model entails the conceptual analysis. Factor analysis provides a means to assess the conceptual validity of the questionnaire items which make up the scales by examining the extent to which they seem to be measuring the same concepts or variables (Bryman and Cramer, 1999).

Two types of factor analysis can be distinguished: *exploratory* factor analysis and *confirmatory* factor analysis. While both types aim to examine the extent to

---

[47] On the basis of these results, the relevant items for 'attitude' and 'social pressure', respectively, were summed to form indices of 'attitude' and 'social pressure' to be used in the subsequent analyses. However, the correlation coefficients for the two items that were used as 'direct' measures of 'perceived control' (.1730 and .1751 for the 'present' and the 'near future' measures respectively) were regarded too low for the scales to be usable. After considering the correlation of both items (for the 'present') individually with the scales of belief-based measures of 'perceived control', it was decided to use as a 'direct' measure of 'perceived control' the item '*After considering our organisation's technological ... spatial data sharing is likely to be easy/difficult*' which had higher correlations, most of which were highly significant.

which variables are measuring distinct underlying concepts, in *exploratory* factor analysis, the number of factors (or concepts) is unknown. The relationships between variables are examined but the data are not tested against a particular model. It can be used, for example, to reduce the number of variables to a number of more manageable and meaningful composite variables (Morgan and Griego, 1998).

In contrast, *confirmatory* factor analysis is used to confirm a hypothesised model. The solution of the factor analysis is compared against a given one, for example, against that suggested by an underlying theory.

The outcome of factor analysis indicates the number of underlying factors (or concepts) that were extracted and the loading of each variable on one or more of the factors. The factor loading expresses the relationship between each variable and the underlying factor or concept (Bryman and Cramer, 1999). The loadings are correlation coefficients of each variable with the factor and range from -1.0 to +1.0.

In confirmatory factor analysis, the loading of each variable on a particular factor(s) is compared to the expected solution. If the variables that load high on one factor fit together conceptually, the underlying concept can be named. These interpretations of the outcomes establish the extent to which the results confirm or disconfirm the hypothesised structure.

In this research, the *principal component extraction* method in SPSS was used for confirmatory factor analysis. The number of factors to be extracted was specified to be three in accordance with the underlying model of spatial data sharing which consists of three main components ('attitude', 'social pressure', and 'perceived control'). A rotated solution of factor analysis output was employed in order to increase the interpretability of the factors since the rotation maximises the loadings of some of the variables and thereby assists in identifying the underlying concept (Bryman and Cramer, 1999). The most common method to increase the interpretability of the factors is *Varimax rotation* (with Kaiser Normalisation in SPSS) (Brosius, 1998).

In order to confirm the structure of the model, confirmatory factor analysis was carried out at two levels. The first level of conceptual verification was the grouping of the direct measures of the domains into each of the three main components: 'attitude', 'social pressure' and 'perceived control'. The second level of verification was the grouping of the beliefs into the three components.

Conceptual Validation Using the Domains

The validation of the main concepts using the domains was carried out using confirmatory factor analysis of the direct measures at the domain-level. Table 5 shows the factor loadings for the direct measures at the domain level.

Table 5　　Results of confirmatory factor analysis at domain level

Direct measure at domain level	Factor loadings Component		
	1 Social pressure	2 Attitude	3 Perceived control
Attitude			
Resource outcomes		.758	
Organisational activities		.549	
Strategic position outcomes		.761	
Social outcomes	.587		
Social pressure			
GIS community pressure	.719		
Market pressure	.826		
Institutional pressure	.782		
Organisational pressure	.621		
Moral norm	.511	.485	
Perceived control			
Technical skills			.769
Interpersonal skills			.765
Resource outcomes			.764
Spatial data position	.557		.405
Finding sharing partners *			
Opportunities	-.513		

Note:　Extraction Method: Principal Component Analysis.
　　　　Rotation Method: Varimax with Kaiser Normalization.[48]
　　　　* There was no question to address 'Finding sharing partners' at the domain level in the questionnaire.
　　　　Only factor loadings above .4 are listed in the table.[49]

The factor loading expresses the relationship between each measure and the underlying factor or concept (Bryman and Cramer, 1999). The clustering of domains on one factor suggests that these are addressing the same underlying concept. As shown in Table 5, the 'social pressure' domains load mainly on factor 1, the 'attitude' domains on factor 2, and the 'perceived control' domains load mainly on factor 3. This grouping of the different domains is in accordance with the three main components ('attitude', 'social pressure' and 'perceived control') of the model of spatial data sharing. This means that the analysis indicates significant loadings for each item on its hypothesised construct.

The exceptions were the domains of social outcomes, moral norms and opportunities. The social outcomes domain loads on the 'social pressure' factor rather than 'attitude', and the moral norms domain loads both, on the 'attitude' and

---

[48] Rotated solution of factor analysis output was used in order to increase the interpretability of the factors since the rotation maximises the loadings of some of the variables (Bryman and Cramer, 1999).

[49] Conventionally, low variable loadings on a factor (in this case, less than .4) are omitted from consideration because they are not very important (Bryman and Cramer, 1999).

'social pressure' factors. The beliefs grouped in these two domains are concerned with morality. As discussed in Chapter 3, the position of morality within the TPB is unsettled in the literature. Therefore, the model of the willingness to share included these two domains to address morality under both, 'attitude' and 'social pressure'. Since the domains refer to very similar beliefs, the cross-loading of the social outcomes and moral norms domains is not unexpected. The negative loading of the opportunity domain shows that it is opposed to factor 1 ('social pressure'). This makes intuitive sense since this domain assesses opportunity creation and is thus conceptually opposed to the expectations and pressures to share assessed by factor 1.

In summary, using the empirical data it was possible to confirm that the domain level questions tapped the three concepts 'attitude', 'social pressure' and 'perceived control' as suggested by the model of spatial data sharing. This means that these questionnaire items are valid by capturing the distinct conceptual aspects they were supposed to capture.

Conceptual Validation Using the Beliefs

The second level of verification was to test the grouping of the different types of combined beliefs (behavioural beliefs with outcome evaluations (*be*), normative beliefs with motivation to comply (*nm*), and control beliefs with perceived control over a skill or resource (*cp*)) into three distinct factors. This grouping was expected to correspond to the three components, 'attitude', 'social pressure' and 'perceived control', of the model of spatial data sharing.

Table 6 shows the factor loadings for the beliefs belonging to 'social pressure' which load predominantly on the first factor. Detailed results of the confirmatory factor analysis for the beliefs belonging to 'attitude' and those belonging to 'perceived control' are presented in Table 7 and Table 8. The factor loadings for the beliefs belonging to 'attitude' load predominantly on the second factor and those for beliefs belonging to 'perceived control' load predominantly on the third factor.

Table 6   Results of confirmatory factor analysis for social pressure beliefs

Construct / items (n = 112)	1 Social pressure	2 Attitude	3 Perceived control
GIS community pressure			
GIS departments in local authorities	.629		
GIS departments in provincial government	.743		
GIS departments in national government	.740		
GIS departments in para-statal organisations	.712		
GIS users in the private sector	.693		
NGO GIS users	.752		
GIS users in academic research institutions	.601		
Market pressure			
Commercial spatial data brokers	.544		
Public spatial data providers	.681		
Private spatial data providers	.622		
GIS suppliers	.662		
Institutional pressure			
NSIF	.487	.352	
Spatial data agreements	.580	.397	
Politicians	.529		
Organisational pressure			
Other departments in the organisation	.556		
Management of the organisation	.597		
Champions for spatial data sharing	.611		
Organisational goals / mission		.427	
Moral Norms			
Integrated development planning	.519	.439	
Society at large	.457	.370	

Note: Extraction Method: Principal Component Analysis.
          Rotation Method: Varimax with Kaiser Normalization.
          Only factor loadings above .3 are listed in the table.

Table 7   Results of confirmatory factor analysis for attitude beliefs

Construct / items (n = 112)	Component 1 Social pressure	Component 2 Attitude	Component 3 Perceived cotnrol
Attitude			
Resource outcomes			
Cost implications			
Time			
Staff			
Data storage			
Administrative efforts			
Expenses			
Savings implications			
Time		.564	
Staff		.584	
Data storage		.490	
Administrative efforts		.596	
Expenses		.582	
Spatial data Outcomes			
Quantity of spatial data		.586	
Introduction of standards		.489	
Quality improvements of spatial data			
Introduction of errors and gaps in spatial data			
Organisational activities			
Focus on core activities	.433		
Usefulness of GIS	.446	.476	
Quality of decision making		.543	
Strategic position outcomes			
Loss of control of spatial data			
No copyright		.425	
Unregulated access		.552	
Use of spatial data not owned		.413	
No enforcement of rules		.417	
Knowledge creation outcomes			
Combination of spatial data	.352	.598	
Access to codified ideas		.476	
Shared experiences		.383	
Trigger dialogue		.442	
Inter-organisational relations outcomes			
Increased interdependence	.401	.431	
Redistribution of influence	.421	.408	
Social outcomes			
Integrated development planning	.422	.496	
Benefits to society at large	.450	.476	

Note: Extraction Method: Principal Component Analysis. Rotation Method: Varimax with Kaiser Normalization. Only factor loadings above .3 are listed in the table.

Table 8   Results of confirmatory factor analysis for perceived control beliefs

Construct / items (n = 112)	Component 1 Social pressure	2 Attitude	3 Perceived control
Perceived control			
Technical skills			
Spatial data skills			
Assessing spatial data quality			.424
Handling different formats of spatial data			.382
Mastering different standards			.543
Selecting spatial data from database			.461
Integrating spatial data from diverse sources			.364
Metadata skills			
Interpreting Metadata		.377	.505
Using Metadata interfaces / catalogues		.472	.576
Capturing Metadata			.536
Applying Metadata standards			.612
Maintaining and updating Metadata			.657
General IT skills			
Database administration skills			.493
Using the Internet to locate spatial data sources			.452
Using the Internet to distribute spatial data sources			.417
Transferring spatial data to / from different media			.411
Interpersonal skills			
Networking and collaboration skills			
Establishing and fostering a network of contacts			.485
Identifying and attending meetings			.551
Keeping a finger on the pulse of a network			.665
Collaborating with others		.369	.515
Interdisciplinary teamwork			.521
Negotiation skills			
Win-win situation			
Pricing of spatial data			.357
Data ownership agreements			.385
Liability agreements			.575
Past experience with sharing			
Heavy past involvement	.387		
Bad past experiences with sharing			
Lack of clear overview of consequences of sharing			

Table 8 continued

Construct / items	Component		
(n = 112)	1	2	3
	Social pressure	Attitude	Perceived control
Resource control			
Staff			.431
Time			
Funding			.427
Organisational guidelines			.542
Spatial data position			
Importance of spatial data			
Importance of other's spatial data for own organisation		-.375	
Importance of own spatial data for other organisations	.507		
Dependence on spatial data			
No copyright			
Unregulated access			
Use of spatial data not owned			
No enforcement of rules			
Control over spatial data			
Copyright		-.432	.416
Regulated access		-.463	.455
Own spatial data not used by other organisations			.500
Enforcement of rules			.663
Availability and stability of SD supplies			
... of required spatial data			.412
... of provided spatial data			
Spatial data self-sufficiency			
Self-sufficiency		-.361	
Spatial data position	.447		
Finding sharing partners			
Willing, responsive			
Reliable			.423
Compatible purpose of application			
'Organisational fit'			.457
Opportunities			
NSIF awareness creation		-.503	
Fora organised by the NSIF		-.469	-.393
Policy development		-.524	
Standards alignment		-.509	
South African core data set identification		-.576	
On-line clearinghouse		-.525	

Note: Extraction Method: Principal Component Analysis. Rotation Method: Varimax with Kaiser Normalization.a Rotation converged in 8 iterations.

The way in which the three types of combined beliefs loaded separately on the three factors in the analysis corresponds to the conceptual distinction between 'attitude', 'social pressure', and 'perceived control'. The empirical data therefore confirmed the conceptual basis of the model which was that the three different types of beliefs would capture the three different conceptual components of the model of spatial data sharing.

Construct Validation of the Model

The third stage of the validation procedure involved testing the adequacy of the proposed hypotheses concerning the structural relationships envisaged by the model.

Relationship between Direct Measures of Main Components and Sums of Beliefs

Beliefs are postulated to provide the basis for the formation of 'attitude', 'social pressure', and 'perceived control'. In the results of the survey, direct measures of the three main components of the model - 'attitude', 'social pressure' and 'perceived control' - should therefore be related to the corresponding sums of combined behavioural, normative and control beliefs. The basis for assessing the relationship between each pair of variables was bivariate correlation analysis to gauge the strength of the relationship. The most common measure of correlation is Pearson's r (Bryman and Cramer, 1999) which usually has been employed to test the assumption of the TPB that significant associations exist between direct and belief-based measures (Ajzen, 1985; Giles and Cairns, 1995; Terry and O'Leary, 1995). Each pair of direct and belief-based measure should be positively correlated as stated in the hypotheses about these relationships presented in Chapter 5.

It was hypothesised that 'attitude' and the sum of the combined behavioural beliefs (Hypothesis 1) are related. In order to test Hypothesis 1, the correlation between the direct measure of 'attitude' (A) towards spatial data sharing and the sum of all combined behavioural beliefs was assessed (see Table 9). Since a significant correlation exists, it can be concluded that the hypothesised relationship between 'attitude' and the sum of the behavioural beliefs is confirmed.[50]

---

[50] Any significance level is arbitrary. Following conventional practice, the .05 cut-off point for significance levels is used throughout this study. Where higher levels of significance were attained (e.g. p<.01 or p<.000), these are indicated.

Table 9   **Correlation between main components and sum of beliefs**

No.	Direct measure of main component	Sum of beliefs	Pearson correlation coefficient (r)
		Hypotheses	
1 Attitude	A $\propto$	$\displaystyle\sum_{i=1}^{29} b_i e_i$	.478**
2 Social pressure	PSP $\propto$	$\displaystyle\sum_{i=1}^{20} n_i m_i$	.617**
3 Perceived control	PBC $\propto$	$\displaystyle\sum_{i=1}^{56} c_i p_i$	.270**

**. Correlation is significant at the 0.01 level (2-tailed)
(n = 112)

'Social pressure' was hypothesised to be related to the sum of combined normative beliefs, as stated in Hypothesis 2. Addressing Hypothesis 2 is the correlation between the direct measure of 'social pressure' (*PSP*) to engage in spatial data sharing and the sum of all the combined normative beliefs (see Table 9). This correlation is significant and it can be concluded that the hypothesised relationship between 'social pressure' and the sum of the combined normative beliefs is confirmed.

'Perceived control' is hypothesised to be related to the sum of combined control beliefs (Hypothesis 3). Responding to Hypothesis 3 is the correlation between the direct measure of 'perceived control' (*PBC*) over spatial data sharing and the sum of all combined control beliefs (see Table 9). As this correlation is significant, it can be concluded that the hypothesised relationship between 'perceived control' and the sum of the combined control beliefs is confirmed.

This section has confirmed that the relationships between the direct measures of the main components ('attitude', 'social pressure', and 'perceived control') and the respective sums of underlying beliefs exist in the empirical data collected in the survey. This completes the first step in the construct validation of the proposed model of the willingness to engage in spatial data sharing across organisational boundaries.

Explanation of Variables by Independent Variables

A second aspect of construct validation concerns the explanatory power of the model of the willingness to share spatial data, as stated in the hypotheses developed in Chapter 5. 'Attitude', 'social pressure' and 'perceived control' were hypothesised to explain the willingness to share (see Hypothesis at Level 1, Table

10). Since beliefs are assumed to provide the basis for 'attitude', 'social pressure' and 'perceived control', measures of the beliefs should serve to predict or explain each of those components (see Hypotheses at Level 2, Table 10).

Table 10   Validation of explanatory relations between variables

No.	Hypotheses	R	F	Significance p
	Level 1			
4	W = W (A, PSP, PBC)	.678	30.611	.000
	Level 2			
5	A = A (Σcosts, Σbenefits, Σspatial data outcomes, Σorganisational activities, Σloss of control, Σknowledge creation, Σinter-org relations, Σsocial outcomes)	.581	6.550	.000
6	PSP = PSP (ΣGIS community, Σmarket, Σinstitutional, Σorganisational pressure, Σmoral norms)	.631	14.030	.000
7	PBC = PBC (Σspatial data, Σmetadata, Σitskills, Σnetworking, Σnegotiation, Σpast experience, Σresource control, Σimportance of internal data, Σcontrol aspects, Σaltern. sources1, Σimportance of external data, Σdependence aspects, Σaltern. sources2, Σself sufficiency, Σsharing partners, Σopportunities)	.570	2.851	.001

Multiple regression can be used to predict a dependent variable from several independent variables and was used to validate the hypotheses regarding the explanatory power of the model. The following aspects were considered as estimates of how well the regression equations fit the data:

- the squared multiple regression coefficient (R^2) is the measure of how well the dependent variable can be predicted knowing only the independent variables;

- the standard error of the estimate should not be much greater than one unit of measurement;
- inspection of the *F* test. It tests whether all the coefficients of the independent variables are zero in the population from which the sample was taken. If the *F* test is not significant this would imply that the independent variables add nothing to the explanation of the dependent variable. If the test is significant, it can be concluded that at least one of the coefficients is not zero;
- inspection of the *t* tests for each coefficient. The t value is the test of statistical significance of individual regression coefficients. A significance level (*p* value) less than 0.05 provides strong evidence that the coefficient is not zero, so that the coefficient can be accepted;
- inspection of tolerance statistics to diagnose multicollinearity. Ranging from 0 to 1, values below .40 indicate danger of multicollinearity;
- inspection of residuals to check that they are normally distributed.

The 'enter' method was used to consider all the variables at the same time. Regression analysis asumes that a linear relationship exists between the dependent and the independent variables. The relationship may be explained formally by a different mathematical function but, for the purpose of this research, it is sufficient to establish that a relationship of dependence between the variables exists via the fit of the data to a linear function.

The main concern of this analysis was to verify the relationship between the dependent and the independent variables. A significant *F* test result is sufficient to indicate that, overall, the regression coefficients are not equal to zero and that the independent variables are contributing to the prediction of the dependent variable (Allison, 1999).[51] At both explanatory levels, the explanatory power of the model has been verified. At the first level, the three components 'attitude', 'social pressure' and 'perceived control' do appear to explain the willingness to share (Hypothesis 4), as indicated by the significant *F* test ($p < .000$). At the second level, the scales of combined behavioural beliefs also appear to explain 'attitude' (Hypothesis 5), the scales of combined normative beliefs seem to contribute to the explanation of 'social pressure' (Hypothesis 6), and the scales of combined control beliefs appear to explain 'perceived control' (Hypothesis 7), again indicated by the significant *F* test results.

More detailed results of the regression analysis are presented in the following subsections.

---

[51] The relative importance of the contribution of each independent variable to explain the dependent variable is examined in the second part of the analysis (see Chapter 7).

Regression Results: Intention - Main Components (Present)

Table 11   Overview of variables

Variables Entered/Removed[b]

Model	Variables Entered	Variables Removed	Method
1	pbc1, att1, psp1[a]	.	Enter

a. All requested variables entered.

b. Dependent Variable: INT3.1

Table 12   Model summary

Model Summary[b]

Model	R	R Square	Adjusted R Square	Std. Error of the Estimate
1	.678[a]	.460	.445	1.11

a. Predictors: (Constant), pbc1, att1, psp1

b. Dependent Variable: INT3.1

Table 13   Anova results

ANOVA[b]

Model		Sum of Squares	df	Mean Square	F	Sig.
1	Regression	112.637	3	37.546	30.611	.000[a]
	Residual	132.467	108	1.227		
	Total	245.104	111			

a. Predictors: (Constant), pbc1, att1, psp1

b. Dependent Variable: INT3.1

Table 14 Coefficients results

Coefficients[a]

Model		Unstandardized Coefficients		Standardized Coefficients	t	Sig.	Collinearity Statistics	
		B	Std. Error	Beta			Tolerance	VIF
1	(Constant)	-7.38E-03	.557		-.013	.989		
	att1	.346	.119	.255	2.901	.005	.650	1.539
	psp1	.516	.101	.462	5.104	.000	.612	1.635
	pbc1	.104	.093	.084	1.121	.265	.889	1.125

a. Dependent Variable: INT3.1

Regression Results: Intention - Main Components (Near Future)

Table 15 Overview of variables

Variables Entered/Removed[b]

Model	Variables Entered	Variables Removed	Method
1	PBC1.2, SUMATT2, SUMPSP2[a]	.	Enter

a. All requested variables entered.

b. Dependent Variable: INT3.2

Table 16 Model summary

Model Summary

Model	R	R Square	Adjusted R Square	Std. Error of the Estimate
1	.709[a]	.503	.489	.99

a. Predictors: (Constant), PBC1.2, SUMATT2, SUMPSP2

Table 17   Anova results

ANOVA[b]						
Model	Sum of Squares	df	Mean Square	F	Sig.	
1 Regression	106.594	3	35.531	36.455	.000[a]	
Residual	105.263	108	.975			
Total	211.857	111				

a.   Predictors: (Constant), PBC1.2, SUMATT2, SUMPSP2

b.   Dependent Variable: INT3.2

Table 18   Coefficents results

Coefficients[a]						
		Unstandardized Coefficients		Standardized Coefficients		
Model		B	Std. Error	Beta	t	Sig.
1	(Constant)	.340	.575		.591	.556
	SUMATT2	.421	.116	.324	3.628	.000
	SUMPSP2	.463	.101	.434	4.600	.000
	PBC1.2	5.271E-02	.093	.043	.564	.574

a. Dependent Variable: INT3.2

Regression Results: Attitude – Beliefs

Table 19   Overview of variables

Variables Entered/Removed[b]

Model	Variables Entered	Variables Removed	Method
1	COSTSCAL, BENSCALE, LOSSCALE, INTERORG, SDSCALE, KNWLSCAL, WQSCALE, SOSCALE [a]	.	Enter

a. All requested variables entered.

b. Dependent Variable: att1

Table 20   Model summary

Model Summary[b]

Model	R	R Square	Adjusted R Square	Std. Error of the Estimate
1	.581[a]	.337	.286	.924

a. Predictors: (Constant), SOSCALE, COSTSCAL, BENSCALE, LOSSCALE, INTERORG, SDSCALE, KNWLSCAL, WQSCALE

b. Dependent Variable: att1

Table 21 Anova results

<table>
<tr><td colspan="7" align="center">ANOVA[b]</td></tr>
<tr><td>Model</td><td>Sum of Squares</td><td>df</td><td>Mean Square</td><td>F</td><td>Sig.</td></tr>
<tr><td>1   Regression</td><td>44.769</td><td>8</td><td>5.596</td><td>6.550</td><td>.000[a]</td></tr>
<tr><td>Residual</td><td>88.002</td><td>103</td><td>.854</td><td></td><td></td></tr>
<tr><td>Total</td><td>132.771</td><td>111</td><td></td><td></td><td></td></tr>
</table>

a. Predictors: (Constant), SOSCALE, COSTSCAL, BENSCALE, LOSSCALE, INTERORG, SDSCALE, KNWLSCAL, WQSCALE

b. Dependent Variable: att1

Table 22 Coefficients results

<table>
<tr><td colspan="9" align="center">Coefficients[a]</td></tr>
<tr><td rowspan="2"></td><td colspan="2">Unstandardized Coefficients</td><td>Standardized Coefficients</td><td rowspan="2">t</td><td rowspan="2">Sig.</td><td colspan="2">Collinearity Statistics</td></tr>
<tr><td>Model</td><td>B</td><td>Std. Error</td><td>Beta</td><td></td><td></td><td>Tolerance</td><td>VIF</td></tr>
<tr><td>1   (Constant)</td><td>1.656</td><td>.745</td><td></td><td>2.222</td><td>.028</td><td></td><td></td></tr>
<tr><td>COSTSCAL</td><td>-2.45E-03</td><td>.003</td><td>-.061</td><td>-.726</td><td>.470</td><td>.910</td><td>1.098</td></tr>
<tr><td>BENSCALE</td><td>3.947E-03</td><td>.002</td><td>.190</td><td>1.961</td><td>.053</td><td>.682</td><td>1.465</td></tr>
<tr><td>SDSCALE</td><td>6.053E-03</td><td>.007</td><td>.102</td><td>.928</td><td>.356</td><td>.532</td><td>1.880</td></tr>
<tr><td>WQSCALE</td><td>1.378E-02</td><td>.052</td><td>.030</td><td>.267</td><td>.790</td><td>.522</td><td>1.915</td></tr>
<tr><td>LOSSCALE</td><td>1.657E-03</td><td>.002</td><td>.065</td><td>.706</td><td>.482</td><td>.772</td><td>1.296</td></tr>
<tr><td>KNWLSCAL</td><td>9.015E-03</td><td>.004</td><td>.279</td><td>2.549</td><td>.012</td><td>.537</td><td>1.861</td></tr>
<tr><td>INTERORG</td><td>-8.64E-03</td><td>.006</td><td>-.170</td><td>-1.558</td><td>.122</td><td>.543</td><td>1.843</td></tr>
<tr><td>SOSCALE</td><td>.113</td><td>.049</td><td>.248</td><td>2.310</td><td>.023</td><td>.557</td><td>1.794</td></tr>
</table>

a. Dependent Variable: att1

Regression Results: Perceived Social Pressure - Beliefs

Table 23   Overview of variables

Variables Entered/Removed[b]

Model	Variables Entered	Variables Removed	Method
1	MNPSCALE, MPSCALE, OPSCALE, INSPSCAL, CPSCALE[a]	.	Enter

a. All requested variables entered.

b. Dependent Variable: psp1

Table 24   Model summary

Model Summary[b]

Model	R	R Square	Adjusted R Square	Std. Error of the Estimate
1	.631[a]	.398	.370	1.054

a. Predictors: (Constant), MNPSCALE, MPSCALE, OPSCALE, INSPSCAL, CPSCALE

b. Dependent Variable: psp1

Table 25 Anova results

ANOVA[b]						
Model		Sum of Squares	df	Mean Square	F	Sig.
1	Regression	77.997	5	15.599	14.030	.000[a]
	Residual	117.857	106	1.112		
	Total	195.854	111			

a. Predictors: (Constant), MNPSCALE, MPSCALE, OPSCALE, INSPSCAL, CPSCALE
b. Dependent Variable: psp1

Table 26 Coefficients results

Coefficients[a]								
		Unstandardized Coefficients		Standardized Coefficients			Collinearity Statistics	
Model		B	Std. Error	Beta	t	Sig.	Tolerance	VIF
1	(Constant)	1.807	.374		4.835	.000		
	CPSCALE	4.229E-03	.002	.236	2.136	.035	.463	2.160
	MPSCALE	2.141E-03	.003	.063	.612	.542	.537	1.864
	INSPSCAL	4.618E-03	.005	.101	.916	.362	.470	2.128
	OPSCALE	1.057E-02	.004	.252	2.555	.012	.582	1.719
	MNPSCALE	8.195E-03	.006	.134	1.411	.161	.634	1.577

a. Dependent Variable: psp1

Regression Results: Perceived Behavioural Control – Beliefs

Table 27   Overview of variables

Variables Entered/Removed[b]

Model	Variables Entered	Variables Removed	Method
1	DD1.2, OPPSCALE, DEPSCAL1, FSHP, CONTRSC2, TSSD, DD1.1, ISNEG, ISEXP, CONTRSC1, TSIT, RCONTR, TSMD, DEPSCAL2, DEPSCAL3, ISNETW [a]		Enter

a.   All requested variables entered

b.   Dependent Variable: pbc 1

Table 28   Model summary

Model Summary[b]

Model	R	R Square	Adjusted R Square	Std. Error of the Estimate
1	.570[a]	.324	.211	.849

a. Predictors: (Constant), DD1.2, OPPSCALE, DEPSCAL1, FSHP, CONTRSC2, TSSD, DD1.1, ISNEG, ISEXP, CONTRSC1, TSIT, RCONTR, TSMD, DEPSCAL2, DEPSCAL3, ISNETW

b. Dependent Variable: pbc1

Table 29   Anova results

		ANOVA[b]				
Model		Sum of Squares	df	Mean Square	F	Sig.
1	Regression	32.909	16	2.057	2.851	.001[a]
	Residual	68.528	95	.721		
	Total	101.437	111			

a. Predictors: (Constant), DD1.2, OPPSCALE, DEPSCAL1, FSHP, CONTRSC2, TSSD, DD1.1, ISNEG, ISEXP, CONTRSC1, TSIT, RCONTR, TSMD, DEPSCAL2, DEPSCAL3, ISNETW

b. Dependent Variable: pbc1

Table 30   Coefficients results

Coefficients [a]

		Unstandardized Coefficients		Standardized Coefficients			Collinearity Statistics	
Model		B	Std. Error	Beta	t	Sig.	Tolerance	VIF
1	(Constant)	4.566	.750		6.086	.000		
	TSSD	-8.86E-04	.002	-.038	-.360	.720	.639	1.564
	TSMD	5.716E-03	.002	.301	2.607	.011	.534	1.872
	TSIT	6.221E-04	.003	.022	.185	.854	.486	2.057
	ISNETW	-2.04E-03	.003	-.094	-.769	.444	.476	2.101
	ISNEG	6.762E-03	.003	.256	2.478	.015	.666	1.501
	ISEXP	-2.25E-02	.028	-.082	-.815	.417	.704	1.420
	RCONTR	4.853E-03	.003	.173	1.542	.126	.568	1.759
	DEPSCAL1	-2.71E-03	.003	-.083	-.835	.406	.728	1.374
	DEPSCAL2	-2.11E-02	.036	-.068	-.582	.562	.525	1.905
	DEPSCAL3	4.852E-03	.038	.015	.128	.898	.527	1.898
	CONTRSC1	-1.04E-02	.002	-.474	-4.433	.000	.621	1.609
	CONTRSC2	-3.10E-03	.036	-.010	-.087	.931	.571	1.752
	FSHP	-2.72E-03	.003	-.093	-.937	.351	.716	1.396
	OPPSCALE	-4.56E-04	.012	-.004	-.039	.969	.820	1.220
	DD1.1	.124	.056	.237	2.210	.030	.618	1.617
	DD1.2	.901E-02	.075	.072	.652	.516	.584	1.713

a. Dependent Variable: pbc1

Examination of NSIF-related Items

This appendix to Chapter 7 contains the results of the correlation analysis and the factor analysis carried out to examine the relationship between items in the questionnaire that referred to the NSIF directly and those that referred to GIS national government departments. The aim was to assess the extent to which items referring to GIS national government departments indirectly included the NSIF.

Results of Correlation Analysis

Table 1   **Correlations of GIS national government department items with NSIF items**

Correlations		CP1.3	CP2.3	INSP1.1	INSP2.1
CP1.3	Pearson Correlation	1,000	,541**	,536**	,382**
	Sig. (2-tailed)	,	,000	,000	,000
	N	112	112	112	112
CP2.3	Pearson Correlation	,541**	1,000	,381**	,493**
	Sig. (2-tailed)	,000	,	,000	,000
	N	112	112	112	112
INSP1.1	Pearson Correlation	,536**	,381**	1,000	,464**
	Sig. (2-tailed)	,000	,000	,	,000
	N	112	112	112	112
INSP2.1	Pearson Correlation	,382**	,493**	,464**	1,000
	Sig. (2-tailed)	,000	,000	,000	,
	N	112	112	112	112

· Correlation is significant at the 0.01 level (2-tailed).

Note:
CP1.3 :   pressure from GIS national government departments with respect to spatial data sharing
CP2.3 :   compliance with expectations from GIS national government departments
INSP1.1 :   pressure from the NSIF with respect to spatial data sharing
INSP2.1 :   compliance with NSIF expectations

The correlations of CP1.3 and CP2.3 with INSP1.1 and INSP2.1 shown in Table 1 are significant.

Results of Factor Analysis

Table 2   Factor analysis of GIS national government department items and NSIF items

Component Matrix[a]

	Component 1
CP1.3	,800
CP2.3	,782
INSP1.1	,767
INSP2.1	,748

Extraction Method: Principal Component Analysis.

a. 1 components extracted.

The result of the factor analysis in Table 2 shows that all four items are correlated strongly with the same underlying factor.

Analysis of Differences in Means of Intention Groups for Determinant Beliefs

This appendix contains the results of difference in mean testing for the determinants of 'attitude' and 'social pressure' at the belief level. These results are discussed in detail in Chapter 7.

For this analysis, the respondents were divided into three groups (*willing, undecided, unwilling*) according to their responses regarding their organisation's willingness to engage in spatial data sharing. The purpose was to identify any significant differences between the means of these three groups for each of the determinant beliefs.

Perceived Pressure from GIS Community

The result of the Levene test is displayed in Table 1. This shows that the Levene test of homogeneity of variances among groups was not significant for any of the GIS community referents so that the assumption of equal variances was not violated.

Table 1   Levene test results (pressure from GIS community sectors)

Test of Homogeneity of Variances

	Levene Statistic	df1	df2	Sig.
CP1.1	1.822	2	109	.167
CP1.2	2.512	2	109	.086
CP1.3	2.513	2	109	.086
CP1.4	.759	2	109	.470
CP1.5	2.028	2	109	.137
CP1.6	.388	2	109	.679
CP1.7	.458	2	109	.634

Table 2   One-Way ANOVA (pressure from GIS community sectors)

ANOVA

		Sum of Squares	df	Mean Square	F	Sig.
CP1.1	Between Groups	11.466	2	5.733	3.350	.039
	Within Groups	186.525	109	1.711		
	Total	197.991	111			
CP1.2	Between Groups	14.472	2	7.236	4.604	.012
	Within Groups	171.304	109	1.572		
	Total	185.777	111			
CP1.3	Between Groups	28.458	2	14.229	9.230	.000
	Within Groups	168.033	109	1.542		
	Total	196.491	111			
CP1.4	Between Groups	4.838	2	2.419	1.342	.266
	Within Groups	196.439	109	1.802		
	Total	201.277	111			
CP1.5	Between Groups	18.484	2	9.242	4.368	.015
	Within Groups	230.623	109	2.116		
	Total	249.107	111			
CP1.6	Between Groups	7.035	2	3.518	2.507	.086
	Within Groups	152.956	109	1.403		
	Total	159.991	111			
CP1.7	Between Groups	5.825	2	2.913	2.404	.095
	Within Groups	132.032	109	1.211		
	Total	137.857	111			

The results of the one-way ANOVA (see Table 2) for each GIS community referent shows that there were overall differences for several of the GIS sectors (CP1.1 - GIS departments in local authorities, CP1.2 - GIS departments in provincial government, CP1.3 - GIS departments in national government, and CP1.5 - GIS users in the private sector).

Motivation to Comply with GIS Community Referents

The Levene test of homogeneity of variances among intention groups was not significant for any of the GIS sectors so that the assumption of equal variances was not violated (shown in Table 3).

Table 3   Levene test results (motivation to comply with GIS community)

Test of Homogeneity of Variances

	Levene Statistic	df1	df2	Sig.
CP2.1	.204	2	109	.816
CP2.2	1.763	2	109	.176
CP2.3	.505	2	109	.605
CP2.4	.003	2	109	.997
CP2.5	.278	2	109	.758
CP2.6	.220	2	109	.803
CP2.7	.627	2	109	.536

Table 4   One-way ANOVA test results (motivation to comply with GIS community)

		Sum of Squares	df	Mean Square	F	Sig.
CP2.1	Between Groups	35.674	2	17.837	6.506	.002
	Within Groups	298.817	109	2.741		
	Total	334.491	111			
CP2.2	Between Groups	41.561	2	20.780	8.024	.001
	Within Groups	282.296	109	2.590		
	Total	323.857	111			
CP2.3	Between Groups	57.538	2	28.769	10.712	.000
	Within Groups	292.739	109	2.686		
	Total	350.277	111			
CP2.4	Between Groups	40.024	2	20.012	8.460	.000
	Within Groups	257.834	109	2.365		
	Total	297.857	111			
CP2.5	Between Groups	21.963	2	10.982	4.473	.014
	Within Groups	267.599	109	2.455		
	Total	289.563	111			
CP2.6	Between Groups	27.815	2	13.908	5.776	.004
	Within Groups	262.435	109	2.408		
	Total	290.250	111			
CP2.7	Between Groups	12.976	2	6.488	2.381	.097
	Within Groups	296.988	109	2.725		
	Total	309.964	111			

ANOVA

The one-way ANOVA could therefore be used to compare the mean motivation to comply with GIS community sectors of the intention groups. The results of this procedure are shown in Table 4. Apart from the motivation to comply with GIS users in academic research institutions (CP1.7), the ANOVA F test was significant for all the other sectors, so that significant differences in the motivation to comply with these sectors were established for the three intention groups.

Mean Perceived Organisational Pressure

The Levene test of homogeneity of variances among groups (shown in Table 5) was significant for OP1.2 (perceived management pressure, p=.012) so that the assumption of equal variances was violated and the Kruskal-Wallis test had to be used for this variable. The means of the other two variables (OP1.1 - pressure from other departments in the organisation and OP1.3 - pressure from sharing 'champions') were compared using a one-way ANOVA.

Table 5   Levene test results (perceived organisational pressure)

Test of Homogeneity of Variances

	Levene Statistic	df1	df2	Sig.
OP1.1	2.205	2	109	.115
OP1.2	4.589	2	109	.012
OP1.3	.442	2	109	.644

The results of the one-way ANOVA (see Table 6) show that there were significant differences between the means of the different intention groups for perceived pressure from other departments within the organisation (OP1.1) and for pressure from champions for spatial data sharing (OP1.3).

Table 6   ANOVA results (perceived organisational pressure)

ANOVA

		Sum of Squares	df	Mean Square	F	Sig.
OP1.1	Between Groups	21.905	2	10.953	7.668	.001
	Within Groups	155.681	109	1.428		
	Total	177.586	111			
OP1.3	Between Groups	14.670	2	7.335	6.640	.002
	Within Groups	120.421	109	1.105		
	Total	135.091	111			

Table 7   Kruskal-Wallis results (perceived management pressure)

Test Statistics[a,b]

	OP1.2
Chi-Square	13.979
df	2
Asymp. Sig.	.001

a. Kruskal Wallis Test
b. Grouping Variable: Intention Groups (3)

As the Kruskal-Wallis test (see Table 7) for perceived management pressure (OP1.2) was significant (p=.001), it can be concluded that there was an overall difference between the mean perceived pressure from management according to three intention groups.

Motivation to Comply with Organisational Referents

The Levene test of homogeneity of variances among groups (shown in Table 8) was significant for OP2.2 (motivation to comply with management, p=.009) and for OP2.3 (motivation to comply with champions for sharing, p=.003) so that the assumption of equal variances was violated and the Kruskal-Wallis test had to be used for these variables. The means of the other two variables (OP2.1 - motivation to comply with other departments in the organisation and OP5 - motivation to comply with organisational goals or mission) were compared using a one-way ANOVA.

Table 8   Levene test results (motivation to comply with organisational referents)

Test of Homogeneity of Variances

	Levene Statistic	df1	df2	Sig.
OP2.1	2.879	2	109	.060
OP2.2	4.920	2	109	.009
OP2.3	6.205	2	109	.003
OP5	.071	2	109	.932

The results of the one-way ANOVA (see Table 9) show that there were no significant differences between the means of the different intention groups for motivation to comply with other departments within the organisation (OP2.1) and with organisational goals or mission (OP5).

Table 9 **ANOVA results (motivation to comply with organisational referents)**

ANOVA

		Sum of Squares	df	Mean Square	F	Sig.
OP2.1	Between Groups	3.093	2	1.546	.863	.425
	Within Groups	195.250	109	1.791		
	Total	198.342	111			
OP5	Between Groups	6.138	2	3.069	2.993	.054
	Within Groups	111.781	109	1.026		
	Total	117.920	111			

Table 10 **Kruskal-Wallis results (motivation to comply with organisational referents)**

Test Statistics[a,b]

	OP2.2	OP2.3
Chi-Square	5.152	8.634
df	2	2
Asymp. Sig.	.076	.013

a. Kruskal Wallis Test

b. Grouping Variable: Intention Groups (3)

As the Kruskal-Wallis test for the motivation to comply with management (OP2.2) was not significant (p=.076), it can be concluded that there was no overall difference between the mean motivation to comply with management according to three intention groups. On the other hand, the Kruskal-Wallis test for motivation to comply with champions for spatial data sharing (OP2.3) was significant (p=.013) and therefore it can be concluded that there was an overall difference between the motivation to comply with champions for sharing according to three intention groups.

Evaluation of Knowledge Creation Outcomes

The Levene test of homogeneity of variances among intention groups was not significant for any of the evaluations of knowledge creation outcomes so that the assumption of equal variances was not violated and the one-way ANOVA test could be used to compare means (shown in Table 11).

Table 11   Levene test results (evaluation of knowledge creation outcomes)

Test of Homogeneity of Variances

	Levene Statistic	df1	df2	Sig.
SPO3.1	3.041	2	109	.052
SPO3.2	1.310	2	109	.274
SPO3.3	1.149	2	109	.321
SPO3.4	.603	2	109	.549

The results of the one-way ANOVA (see Table 12) show that there were significant differences between the means of the different intention groups only for the evaluation of the first knowledge creation outcome (SPO3.1 - combination of spatial data, p=.008).

Table 12   ANOVA results (evaluation of knowledge creation outcomes)

ANOVA

	Sum of Squares	df	Mean Square	F	Sig.
SPO3.1 Between Groups	13.841	2	6.921	5.000	.008
Within Groups	150.873	109	1.384		
Total	164.714	111			
SPO3.2 Between Groups	7.887	2	3.944	2.192	.117
Within Groups	196.077	109	1.799		
Total	203.964	111			
SPO3.3 Between Groups	3.467	2	1.734	1.479	.233
Within Groups	127.809	109	1.173		
Total	131.277	111			
SPO3.4 Between Groups	.454	2	.227	.252	.778
Within Groups	98.108	109	.900		
Total	98.563	111			

Likelihood of Knowledge Creation Outcomes

The Levene test of homogeneity of variances among groups (shown in Table 13) was significant for SPO4.4 (sharing would trigger dialogue, p=.014) so that the assumption of equal variances was violated and the Kruskal-Wallis test had to be used for this variable. The means of the remaining variables (SPO4.1, SPO4.2 and SPO4.3) were compared using a one-way ANOVA.

Table 13   Levene test results (likelihood of knowledge creation outcomes)

Test of Homogeneity of Variances

	Levene Statistic	df1	df2	Sig.
SPO4.1	.860	2	109	.426
SPO4.2	.274	2	109	.761
SPO4.3	1.996	2	109	.141
SPO4.4	4.452	2	109	.014

The results of the one-way ANOVA (see Table 14) show that there were no significant differences between the means of the different intention groups for the likelihood of combination of spatial data (SPO4.1), access to codified ideas (SPO4.2) and shared experiences (SPO4.3).

Table 14   ANOVA results (likelihood of knowledge creation outcomes)

ANOVA

		Sum of Squares	df	Mean Square	F	Sig.
SPO4.1	Between Groups	3.958	2	1.979	2.475	.089
	Within Groups	87.149	109	.800		
	Total	91.107	111			
SPO4.2	Between Groups	1.833	2	.916	1.069	.347
	Within Groups	93.466	109	.857		
	Total	95.299	111			
SPO4.3	Between Groups	6.128	2	3.064	2.537	.084
	Within Groups	131.623	109	1.208		
	Total	137.751	111			

Table 15   Kruskal-Wallis results (likelihood of knowledge creation outcomes)

Test Statistics[a,b]

	SPO4.4
Chi-Square	12.223
df	2
Asymp. Sig.	.002

[a.] Kruskal Wallis Test

[b.] Grouping Variable: Intention Groups (3)

The Kruskal-Wallis test (see Table 15) for the likelihood that sharing would triggering dialogue (SPO4.4) was significant (p=.002) and therefore it can be concluded that there was an overall difference between the means for this variable according to three intention groups.

Likelihood of Social Outcomes

The Levene test of homogeneity of variances among intention groups was not significant for either of the likelihood assessments of social outcomes (see Table 16) so that the assumption of equal variances was not violated and the one-way ANOVA test could be used to compare means.

Table 16   Levene test results (likelihood of social outcomes)

Test of Homogeneity of Variances

	Levene Statistic	df1	df2	Sig.
SO1.1	1.525	2	109	.222
SO1.2	1.776	2	109	.174

The results of the one-way ANOVA (see Table 17) show that there were significant differences between the means of the different intention groups for their assessment of the likelihood of social outcomes to result from spatial data sharing (SO1.1 - integrated development planning and SO1.2 - benefits to society at large).

Table 17   ANOVA results (likelihood of social outcomes)

ANOVA

		Sum of Squares	df	Mean Square	F	Sig.
SO1.1	Between Groups	34.853	2	17.426	11.067	.000
	Within Groups	171.638	109	1.575		
	Total	206.491	111			
SO1.2	Between Groups	19.082	2	9.541	7.656	.001
	Within Groups	135.838	109	1.246		
	Total	154.920	111			

Appendix H

Correlation between Motivation to Comply and Outcome Evaluations

This appendix to Chapter 7 contains the results of the correlation analysis between the motivation to comply with the GIS community sectors and outcome evaluations, and between the motivation to comply with organisational referents and outcome evaluations. The correlation analysis is used to establish whether a significant relationship exists between the 'motivation to comply' with particular referents and the evaluation of the possible outcomes of spatial data sharing. These results are discussed in detail in Chapter 7.

Table 1 Correlation between 'motivation to comply' with the GIS community sectors and outcome evaluations

Outcome evaluations (n = 112)	Pearson correlation coefficient (r) *Motivation to comply with:*						
	GIS dep. in local authorities	GIS dep. in provincial government	GIS dep. in national government	GIS dep. in para-statal orgs.	GIS users in the private sector	NGO GIS users	GIS users in acad.emic / research institutions
Resource outcomes							
Magnitude of cost implications							
Time required to locate spatial data	-.039	.018	-.024	-.086	-.070	-.091	-.005
Trained staff required to exchange and integrate spa tial data	-.078	-.145	-.157	-.095	.016	-.108	-.006
Data storage required for spatial data to be shared	-.119	-.025	-.177	-.145	-.168	-.181	-.139
Administrative efforts for sharing	-.090	-.073	-.172	-.184	-.163	-.174	-.096
Expenses incurred from spatial data sharing	.000	.010	-.060	-.035	.010	-.046	.066
Magnitude of savings implications							
Time saved otherwise needed to capture spatial data in -house	.121	.093	.098	.131	.198*	.206*	.104
Trained staff saved otherwise needed for capturing and maintaining spatial data	.186*	.163	.230*	.239*	.264**	.295**	.094
Data storage saved otherwise needed for divergent/duplicated databases	.064	.071	.200*	.102	.080	.173	.067
Administrative efforts saved otherwise needed for data capture	.198*	.100	.208*	.127	.096	.218*	.093
Expenses saved otherwise needed for data capture	.034	.009	.137	.056	.074	.143	.114
Outcomes for spatial data							
Increased quantity of spatial data available to the organisation	.222*	.127	.119	.094	.107	.093	.112
Introduction of standards agreed with other organisations	.296**	.344**	.262**	.346**	.315**	.244**	.052
Organisational activities							
Focus on core activity	.298**	.390**	.333**	.249**	.162	.250**	.199*
Usefulness of the GIS	.432**	.459**	.425**	.412**	.323**	.380**	.394**
Quality of decision making	.301**	.291**	.316**	.248**	.188*	.313**	.246**

Table 1   continued

Outcome evaluations (n = 112)	Pearson correlation coefficient (r) *Motivation to comply with:*						
	GIS dep. in local authorities	GIS dep. in provincial government	GIS dep. in national government	GIS dep. in para-statal orgs.	GIS users in the private sector	NGO GIS users	GIS users in acad./emic / research institutions
Strategic position outcomes							
Loss of control over spatial data							
Not copyrighting spatial data sets	.055	.105	.076	.021	-.079	-.015	.034
Unregulated access to organisation's spatial data by other organisations	.259**	.286**	.266**	.203*	.109	.180	.162
Use of organisation's spatial data by organisations other than the sharing partners	.370**	.318**	.228*	.259**	.129	.209*	.150
Not enforcing any rules regarding the ownership of, access to and use of the organisation's spatial data	.204*	.193*	.189*	.176	.044	.138	.151
Knowledge creation outcomes							
Combination of new and existing spatial data sets	.287**	.288**	.279**	.307**	.178	.253**	.143
Access to ideas	.210*	.206*	.215*	.188*	.128	.230*	.030
Observe spatial data skills	.054	.086	.072	.059	.072	.104	-.008
Trigger dialogue and collective reflection	.101	.122	.121	.156	.145	.217*	.031
Inter-organisational relations outcomes							
Increased interdependence among organisations	.309**	.364**	.305**	.376**	.383**	.401**	.136
Redistribution of influence among organisations	.293**	.335**	.327**	.341**	.320**	.365**	.071
Social outcomes							
Integrated development planning	.356**	.354**	.340**	.363**	.245**	.368**	.285**
Benefits to society at large	.285**	.339**	.377**	.352**	.176	.279**	.266**

Note: **. Correlation is significant at the 0.01 level (2-tailed)
*. Correlation is significant at the 0.05 level (2-tailed)
For the items under social outcomes, respondents were not asked to *evaluate* them but only to assess the *likelihood*.

Table 2 Correlation between 'motivation to comply' with organisational referents and outcome evaluations

Outcome evaluations of spatial data sharing across organisational boundaries (n = 112)	Pearson correlation coefficient (r) Motivation to comply with:			
	Other departments within the organisation	Management	Champions for spatial data sharing	Organisational goal / mandate
Resource outcomes				
Magnitude of cost implications				
Time required to locate spatial data	-.101	.012	-.225*	.104
Trained staff required to exchange and integrate spatial data	-.194*	-.080	-.139	.186
Data storage required for spatial data to be shared	-.098	.097	.002	-.132
Administrative efforts for sharing	-.127	-.026	-.232*	.049
Expenses incurred from spatial data sharing	-.114	.099	.033	-.045
Magnitude of savings implications				
Time saved otherwise needed to capture spatial data in-house	-.093	.076	.197*	.105
Trained staff saved otherwise needed for capturing and maintaining spatial data	-.071	.056	**.260****	.127
Data storage saved otherwise needed for divergent/duplicated databases	.077	.128	.094	.130
Administrative efforts saved otherwise needed for data capture	-.011	.041	.130	.154
Expenses saved otherwise needed for data capture	.038	.020	.122	.119
Outcomes for spatial data				
Increased quantity of spatial data available to the organisation	.062	.143	.223*	.155
Introduction of standards agreed with other organisations	.043	.128	**.261****	**.269****

Table 2   continued

Outcome evaluations of spatial data sharing across organisational boundaries (n = 112)	Pearson correlation coefficient (r) Motivation to comply with:			
	Other departments within the organisation	Management	Champions for spatial data sharing	Organisational goal / mandate
Organisational activities				
Focus on core activity	.346**	.194*	.224*	.380**
Usefulness of the GIS	.189*	.271**	.230*	.248**
Quality of decision making	.074	.186*	.191*	.213*
Strategic position outcomes				
Loss of control over spatial data				
Not copyrighting spatial data sets	.085	-.077	.121	.064
Unregulated access to organisation's spatial data by other organisations	.108	.078	.226*	.214*
Use of organisation's spatial data by organisations other than the sharing partners	.038	.051	.244**	.187*
Not enforcing any rules regarding the ownership of, access to and use of the organisation's spatial data	.058	-.044	.148	.116
Knowledge creation outcomes				
Combination of new and existing spatial data sets	-.011	.045	.270**	-.005
Access to ideas	.104	-.042	.208*	-.067
Observe spatial data skills	.031	.056	.157	-.037
Trigger dialogue and collective reflection	-.029	-.073	.010	.136
Interorganisational relations outcomes				
Increased interdependence among organisations	-.006	.056	.218*	.083
Redistribution of influence among organisations	.003	.011	.181	.082
Social outcomes				
Integrated development planning	.160	.140	.316**	.139
Benefits to society at large	.183	.159	.316**	.167

Note: **. Correlation is significant at the 0.01 level (2-tailed)
  *. Correlation is significant at the 0.05 level (2-tailed)
For the items under social outcomes, respondents were not asked to evaluate them but only to assess the likelihood.

Bibliography

Aangenbrug, R. T. (1991) 'A Critique of GIS' in D. J. Maguire, M. F. Goodchild, and D. Rhind, (eds) *Geographical Information Systems - Principles and Applications*, vol. 1, Harlow: Longman Scientific & Technical, pp. 101-107.

Abbott, J. (1996) 'The Alignment of GIS in the Public Service in SA', University of Cape Town, Rondebosch, South Africa, Report produced for the RDP Ministry in the Office of the President, 11 May.

Abelson, R. P. and Levi, A. (1985) 'Decision Making and Decision Theory' in G. Lindzey and E. Aronson, (eds) *The Handbook of Social Psychology*, 3rd ed, vol. 1, New York: Random House, pp. 231-309.

AGI (Association for Geographic Information) (1997) 'The Effective Supply and Use of Government held Geographic Information', Association for Geographic Information, London, Report on the AGI/Government programme of round tables, No. 2/97.

Ajzen, I. (1985) 'From Intentions to Actions: A Theory of Planned Behavior' in J. Kuhl and J. Beckmann, (eds) *Action Control - From Cognition to Behavior*, Berlin: Springer-Verlag, pp. 11-39.

Ajzen, I. (1988) *Attitudes, Personality, and Behavior*, Milton Keynes: Open University Press.

Ajzen, I. (1991) 'The Theory of Planned Behaviour', *Organizational Behaviour and Human Decision Processes*, vol. 50, pp.179-211.

Ajzen, I. (1996) 'The Social Psychology of Decision Making' in E. T. Higgins and A. W. Kruglanski, Eds. *Social Psychology - Handbook of Basic Principles*, New York and London: The Guildford Press, pp. 297-325.

Ajzen, I. and Driver, B. L. (1992) 'Application of the Theory of Planned Behavior to Leisure Choice', *Journal of Leisure Research*, vol. 24(3), pp. 207-224.

Ajzen, I. and Fishbein, M. (1980) *Understanding Attitudes and Predicting Social Behavior*, Englewood Cliffs, N.J.: Prentice-Hall, Inc.

Ajzen, I. and Madden, T. J. (1986) 'Prediction of Goal-Directed Behavior: Attitudes, Intentions, and Perceived Behavioral Control', *Journal of Experimental Social Psychology*, vol. 22, pp. 453-474.

Alfelor, R. M. (1995) 'GIS and the Integrated Highway Information System' in H. J. Onsrud and G. Rushton, (eds) *Sharing Geographic Information*, New Brunswick, NJ: Center for Urban Policy Research, pp. 397-412.

Allison, P. D. (1999) *Multiple Regression - A Primer*, Thousand Oaks, CA: Pine Forge Press.

Alter, C. and Hage, J. (1993) *Organisations Working Together*, Newbury Park: Sage Publications.

Antonelli, C. (1995) 'The Diffusion of New Information Technologies and Productivity Growth', *Journal of Evolutionary Economics*, vol. 5, pp.1-17.

Armstrong, M. P. (1993) 'On Automated Geography', *Professional Geography*, vol. 45(4), pp. 440-442.

Aronson, E., Ellsworth, P., Carlsmith, J. M., and Gonzales, M. H. (1990) *Methods of Research in Social Psychology*, 2nd ed, New York: McGraw-Hill.

Azad, B. and Wiggins, L. L. (1995) 'A Proposed Structure for Observing Data Sharing' in H. J. Onsrud and G. Rushton, (eds) *Sharing Geographic Information*, New Brunswick, NJ: Center for Urban Policy Research, pp. 22-43.

Baarck, E. (1986) *The Context of National Information Systems in Developing Countries - India and China in a Comparative Perspective*, Lund, Sweden: Research Policy Institute.

Bacharach, S., Bamberger, P., and Mundell, B. (1995) 'Strategic and Tactical Logics of Decision Justification: Power and Decision Criteria in Organizations', *Human Relations*, vol. 48(5), pp. 467-488.

Bagozzi, R. P. (1986) 'Attitude Formation Under the Theory of Reasoned Action and a Purposeful Behaviour Reformulation', *British Journal of Social Psychology*, vol. 25, pp. 95-107.

Bagozzi, R. P. (1989) 'An Investigation of the Role of Affective and Moral Evaluations in the Purposeful Behaviour Model of Attitude', *British Journal of Social Psychology*, vol. 28, pp. 97-113.

Bagozzi, R. P. and Kimmel, S. K. (1995) 'A Comparison of Leading Theories for the Prediction of Goal-directed Behaviours', *British Journal of Social Psychology*, vol. 34, pp. 437-461.

Bamberger, W. J. (1995) 'Sharing Geographic Information Among Local Government Agencies in the San Diego Region' in H. J. Onsrud and G. Rushton, (eds) *Sharing Geographic Information*, New Brunswick, NJ: Center for Urban Policy Research, pp. 119-137.

Batty, M. (1992) *Sharing Information in Third World Planning Agencies*, National Center for Geographic Information and Analysis, Buffalo, Technical Report No. 92-8, February.

Bernadsen, T. (1992) 'GIS in Developing Countries' in T. Bernadsen, (ed.) *Geographic Information Systems*, Arendal, Norway: Viak IT, pp. 267-276.

Bhyat, M. (1998) 'Development Information Systems GIS', in D. Clarke, E. Gavin, W. Honu, T. Krieg, M. Muller, H.J. Smith, T. Smith, and S. Vorster, (eds) *National Spatial Information Framework Workshop*, proceedings, Sinodale Sentrum, Pretoria, 11 February, pp. 28-29.

Binswanger, H. P. and Ruttan, V. W. (1978) *Induced Innovation - Technology, Institutions, and Development*, Baltimore and London: The Johns Hopkins University Press.

Black, J. (1997) *Maps and Politics*, London: Reaktion Books.

Braga Rodrigues, S. and Hickson, D. J. (1995) 'Success in Decision Making: Different Organizations, Differing Reasons for Success', *Journal of Management Studies*, vol. 32(5), pp. 655-678.

Bryman, A. (1989) *Research Methods and Organization Studies*, London: Routledge.

Bryman, A. and Cramer, D. (1999) *Quantitative Data Analysis with SPSS Release 8 for Windows - A Guide for Social Scientists*, London and New York: Routledge.

Bulmer, M. and Warwick, D. (eds) (1993) *Social Research in Developing Countries - Surveys and Censuses in the Third World*, London: UCL Press Ltd.

Burrough, P. A. and Masser, I. (Eds.) (1998) *European Geographic Information Infrastructures - Opportunities and Pitfalls*, London: Taylor & Francis Ltd.

Burrough, P. A. and McDonnell, R. A. (1998) *Principles of Geographical Information Systems*, Oxford: Oxford University Press.

Calkins, H. W. and Weatherbe, R. (1995) 'Taxonomy of Spatial Data Sharing' in H. J. Onsrud and G. Rushton, (eds) *Sharing Geographic Information*, New Brunswick, NJ: Center for Urban Policy Research, pp. 65-75.

Campbell, H. (1991) 'Organizational Issues in Managing Geographic Information' in I. Masser and M. Blakemore, (eds) *Handling Geographical Information: Methodology and Potential Applications*, Harlow: Longman Scientific & Technical, pp. 259-282.

Campbell, H. (1996) 'A Social Interactionist Perspective on Computer Implementation', *Journal of the American Planning Association*, vol. 62(1), pp. 99-107.

Carnoy, M. (1997) 'The New Information Technology - International Diffusion and its Impact on Employment and Skills', *International Journal of Manpower*, vol. 18(1/2), pp. 119-159.

Cartwright, T. J. (1993) 'Geographic Information Technology as Appropriate Technology for Development' in I. Masser and H. J. Onsrud, (eds) *Diffusion and Use of Geographic Information Technologies*, Dordrecht: Kluwer Academic Publishers, pp. 261-274.

Castle, G. (ed.) (1993) *Profiting from Geographic Information Systems, Fort* Collins, CO: GIS World, Inc.

Christiansen, T., Christ, H., and Hansmann, B. (1997) 'GIS in German Technical Co-operation: The Status Quo in 1996', *GTZ GIS Newsletter* (3), pp. 3-6.

Cialdini, R. B. and Trost, M. R. (1998) 'Social influence: Social Norms, Conformity, and Compliance' in D.T. Gilbert, S.T. Fiske, and G. Lindzey, (eds) *The Handbook of Social Psychology,* 4th ed, vol. 2, pp. 151-192.

CIE (Centre for International Economics) (2000) 'Scoping the Business Case for SDI Development', prepared for GSDI Steering Committee, Centre for International Economics, Canberra & Sydney, March.

Clarke, D. G. (1997) 'Mapping for Reconstruction of South Africa' in D. Rhind, (ed.) *Framework for the World*, Cambridge: GeoInformation International, pp. 48-62.

Clarke, D., Gavin, E., Honu, W., Krieg, T., Muller, M., Smith, H. J., Smith, T., and Vorster, S. (eds) (1998) *National Spatial Information Framework Workshop*, proceedings, Sinodale Sentrum, Pretoria, 11 February.

Conner, M. and Armitage, C. (1998) 'Extending the Theory of Planned Behavior: A Review and Avenues for Further Research', *Journal of Applied Social Psychology*, vol. 28(15), pp. 1429-1464.

Cooke, D. F. (1995) 'Sharing Street Centerline Spatial Databases' in H. J. Onsrud and G. Rushton, (eds) *Sharing Geographic Information*, New Brunswick, NJ: Center for Urban Policy Research, pp. 363-376.

Coombs, R., Walsh, V., and Saviotti, P. (1987) 'The Diffusion of Technological Innovation' in R. Coombs, V. Walsh, and P. Saviotti, (eds) *Economics and Technological Change*, pp. 120-134.

Coppock, J. T. (1991) 'Retrospect and Prospect: a Personal View' in I. Masser and M. Blakemore, (eds) *Handling Geographic Information: Methodology and Potential applications*, Harlow: Longman Scientific & Technical, pp. 285-304.

Coppock, J. T. and Rhind, D. W. (1991) 'The History of GIS' in D. J. Maguire, M. F. Goodchild, and D. W. Rhind, (eds) *Geographical Information Systems - Principles and Applications*, vol. 1, Harlow: Longman Scientific & Technical, pp. 21-43.

Cox, G. W. (1999) 'The Empirical Content of Rational Choice Theory', *Journal of Theoretical Politics*, vol. 11(2), pp. 147-169.

Craig, W. J. (1995) 'Why We Can't Share Data: Institutional Inertia' in H. J. Onsrud and G. Rushton, (eds) Sharing *Geographic Information*, New Brunswick, NJ: Center for Urban Policy Research, pp. 107-118.

Cramer, D. (1998) *Fundamental Statistics for Social Research*, London and New York: Routledge.

Curry, M. R. (1995) 'Geographic Information Systems and the Inevitability of Ethical Inconsistencies' in J. Pickles, (ed.) *Ground Truth - The Social Implications of*

Geographic Information Systems, New York and London: The Guildford Press, pp. 68-87.

Dale, P. and McLaughlin, J. D. (1988) *Land Information Management - An Introduction with Special Reference to Cadastral Problems in Third World Countries*, Oxford: Clarendon Press.

Dangermond, J. (1995) 'Public Data Access: Another Side of GIS Data Sharing' in H. J. Onsrud and G. Rushton, (eds) *Sharing Geographic Information*, New Brunswick, NJ: Center for Urban Policy Research, pp. 331-339.

Dataquest (1996) 'GIS Applications Move into the "Mainstream"', http://gartner5.gartnerweb.com/ (last accessed 19.6.2000).

Dataquest (1999) 'Pushing into the Mainstream', http://gartner5.gartnerweb.com/ (last accessed 19.6.2000).

Davenport, T. H. and Prusak, L. (1998) *Working Knowledge - How Organisations Manage What They Know*, Boston, MA: Harvard Business School Press.

Dawes, R. M. (1998) 'Behavioral Decision Making and Judgement' in D.T. Gilbert, S.T. Fiske, and G. Lindzey, (eds) *The Handbook of Social Psychology*, 4th ed, vol. 1, pp. 497-548.

de Vaus, D. A. (1996) *Surveys in Social Research*, London: UCL Press Limited.

Department of the Environment (1987) *Handling Geographic Information*, Report to the Secretary of State for the Environment of the Committee of Enquiry into the Handling of Geographic Information, London: HMSO.

Dobson, J. E. (1993) 'The Geographic Revolution: A Retrospective on the Age of Automated Geography', *Professional Geographer*, vol. 45(4), pp. 431-439.

Driscoll, R. S. (1992) 'Remote Sensing Provides Data for Developing Countries', *GIS World* (November), pp. 90.

Dueker, K. J. and Vrana, R. (1995) 'Systems Integration: A Reans and a Means for Data Sharing' in H. J. Onsrud and G. Rushton, (eds) *Sharing Geographic Information*, New Brunswick, NJ: Center for Urban Policy Research, pp. 149-171.

Eagly, A. H. and Chaiken, S. (1993) *The Psychology of Attitudes*, Fort Worth: Harcourt Brace College Publishers.

Eagly, A. H. and Chaiken, S. (1998) 'Attitude Structure and Judgement' in D.T. Gilbert, S.T. Fiske, and G. Lindzey, (eds). *The Handbook of Social Psychology*, 4th ed, vol. 1, pp. 269-322.

Ebers, M. (1999) 'The Dynamics of Inter-Organisational Relationships' in S. B. Andrews and D. Knoke, (eds) *Networks in and around Organizations*, vol. 16, Stamford, CT: Jai Press Inc., pp. 31-56.

Edge, D. (ed.) (1994) *Social Studies of Science - An International Review of Research in the Social Dimension of Science and Technology*, vol. 24, London: Sage Publications.

Eisenhardt, K. M. and Zbaracki, M. J. (1992) 'Strategic Decision Making', *Strategic Management Journal*, vol. 13, pp. 17-37.

Elliott, R., Jobber, D., and Sharp, J. (1995) 'Using the Theory of Reasoned Action to Understand Organizational Behaviour: The Role of Belief Salience', *British Journal of Social Psychology*, vol. 34, pp. 161-172.

Epstein, E. F. (1995) 'Control of Public Information' in H. J. Onsrud and G. Rushton, (eds) *Sharing Geographic Information*, New Brunswick, NJ: Center for Urban Policy Research, pp. 307-318.

Esnard, A.-M. and MacDougall, E. B. (1997) 'Common Ground for Integrating Planning Theory and GIS Topics', *Journal of Planning Education and Research*, vol. 17(1), pp. 55-62.

ESRI (Environmental Systems Research Institute Inc.) (1993) *Understanding GIS - The ARC/INFO Method*, Harlow and New York: Longman Scientific & Technical and John Wiley & Sons.

Evans, J. and Ferreira Jr., J. (1995) 'Sharing Spatial Information in an Imperfect World: Interactions between Technical and Organizational Issues' in H. J. Onsrud and G. Rushton, (eds) *Sharing Geographic Information*, New Brunswick, NJ: Center for Urban Policy Research, pp. 448-460.

Evans, J. D. (1997) 'Infrastructures for Sharing Geographic Information Among Environmental Agencies', unpublished DPhil Thesis, Department of Urban Studies and Planning, Boston, MA: Massachusetts Institute of Technology.

Federal Geographic Data Committee (FGDC) (1997) *Framework Introduction and Guide*, Washington, DC.: Federal Geographic Data Committee.

Federal Register (1994) *Coordinating Geographic Data Acquistion and Access: The National Spatial Data Infrastructure*, vol. 59, Federal Register, pp. 17671-17674.

Feldman, J. (1965) 'Organizational Decision Making' in J. March, (ed.) *Handbook of Organizations*, Chicago: Rand McNally, pp. 614-649.

Fishbein, M. and Ajzen, I. (1975) *Belief, Attitude, Intention, and Behavior: An Introduction to Theory and Research*, Reading, MA: Addison-Wesley.

Flowerdew, R. and Green, M. (1991) 'Data Integration: Statistical Methods for Transferring Data between Zonal Systems' in I. Masser and M. Blakemore, (eds) *Handling Geographical Information: Methodology and Potential Applications*, New York: Longman Scientific & Technical, pp. 38-54.

Fourie, H. (1998) 'GI in the SANDF', in Clarke, D., Gavin, E., Honu, W., Krieg, T., Muller, M., Smith, H. J., Smith, T., and Vorster, S., (eds) *National Spatial Information Framework Workshop*, proceedings, Sinodale Sentrum, Pretoria, 11 February, pp. 30-31.

Frank, A.U. (1992) 'Acquiring a Digital Base Map: a Theoretical Investigation into a Form of Sharing Data', *Journal of the Urban and Regional Information Systems Association*, vol. 4(1), pp. 10-23.

Frederick, D. (1995) 'Coordination of Surveying, Mapping and Related Spatial Data Activities' in H. J. Onsrud and G. Rushton, (eds) *Sharing Geographic Information*, New Brunswick, NJ: Center for Urban Policy Research, pp. 355-362.

Freeman, C. (1982) *The Economic of Industrial Innovation*, Second ed, London: Frances Pinter.

Freeman, C. and Soete, L. (1997) *The Economics of Industrial Innovation*, London & Washington: Pinter Publisher.

Galaskiewicz, J. and Zaheer, A. (1999) 'Networks of Competitive Advantage' in S. B. Andrews and D. Knoke, (eds) *Networks in and around Organizations*, vol. 16, Stamford, CT: Jai Press Inc., pp. 237-261.

Gavin, E. (1998) 'Introducing the NSIF', in D. Clarke, E. Gavin, W. Honu, T. Krieg, M. Muller, H.J. Smith, T. Smith, and S. Vorster, (eds) *National Spatial Information Framework Workshop*, proceedings, Sinodale Sentrum, Pretoria, 11 February, pp. 18-19.

Ghiselli, E.E., Campbell, J.P. and Zedeck, S. (1981) *Measurement Theory for Behavioral Sciences*, San Francisco: W.H. Freeman and Company.

Gilbert, D. T., Fiske, S. T., and Lindzey, G. (eds) (1998) *The Handbook of Social Psychology*, Boston, MA: McGraw-Hill, Inc.

Giles, M. and Cairns, E. (1995) 'Blood Donation and Ajzen's Theory of Planned Behaviour: An Examination of Perceived Behavioural Control', *British Journal of Social Psychology*, vol. 34, pp. 173-188.

Goodchild, M. E. (1993) 'Ten Years Ahead: Dobson's Automated Geography in 1993', *Professional Geographer*, vol. 45(4), pp. 444-446.

Goodchild, M. F. (1995a) 'Sharing Imperfect Data' in H. J. Onsrud and G. Rushton, (eds) *Sharing Geographic Information*, New Brunswick, NJ: Center for Urban Policy Research, pp. 413-425.

Goodchild, M. F. (1995b) 'Geographic Information Systems and Geographic Research' in J. Pickles, (ed.) *Ground Truth - The Social Implications of Geographic Information Systems*, New York and London: The Guildford Press, pp. 31-50.

Goodman, P. S. (1993) 'Implementation of New Information Technology' in I. Masser and H. J. Onsrud, (eds) *Diffusion and Use of Geographic Information Technologies*, Dordrecht: Kluwer Academic Publishers, pp. 45-58.

Grandori, A. and Soda, G. (1995) 'Inter-firm Networks: Antecedents, Mechanisms and Forms', *Organization Studies*, vol. 16(2), pp. 183-214.

Granovetter, M. S. (1972) 'The Strength of Weak Ties', *American Journal of Sociology*, vol. 78(6), pp. 1360-1380.

Grube, J. W., Morgan, M., and McGree, S. T. (1986) 'Attitude and Normative Beliefs as Predictors of Smoking Intentions and Behaviours: A Test of Three Models', *British Journal of Social Psychology*, vol. 25, pp. 81-93.

Harrigan, K. R. and Newman, W. H. (1990) 'Bases of Interorganization Co-operation: Propensity, Power, Persistence', *Journal of Management Studies*, vol. 27(4), pp. 417-434.

Harris, T. M., Weiner, D., Warner, T. A., and Levin, R. (1995) 'Pursuing Social Goals Through Participatory Geographic Information Systems' in J. Pickles, (ed.) *Ground Truth - The Social Implications of Geographic Information Systems*, New York and London: The Guildford Press, pp. 196-222.

Harvey, F. and Chrisman, N. (1998) 'Boundary Objects and the Social Construction of GIS technology', *Environment and Planning*, vol. 30, pp. 1683-1694.

Haywood, T. (1995) *Info-Rich - Info-Poor: Access and Exchange in the Global Information Society*, London: Bowker Saur.

Heikkila, E. J. (1998) 'GIS is Dead; Long live GIS!', *American Planning Association Journal*, vol. 64(3), pp. 350-360.

Heiner, R. A. (1988) 'Imperfect Decisions in Organizations', *Journal of Economic Behavior and Organization*, vol. 9, pp. 25-44.

Hogg, M. and Vaughan, G. (1995) *Social Psychology - An Introduction*, London: Prentice Hall.

Howell, J. (1999) 'GIS Implementation at the IEC', presented at *Earth Data Information Systems Conference*, CSIR, Pretoria, 12-14 July.

Howitt, D. and Cramer, D. (1997) *An Introduction to Statistics in Psychology*, London: Prentice Hall.

Hunter, G. (1993) 'Education and Training: Providing the Human Resources of GIS' in G. Castle, (ed.) *Profiting from Geographic Information Systems*, Fort Collins, CO: GIS World, Inc., pp. 350-353.

Jarque, C. M. (1997) 'An Application of New Technologies: the National Geographic Information System of Mexico' in D. Rhind, (ed.) *Framework for the World*, Cambridge: GeoInformation International, pp. 63-70.

Jones, E. E. (1985) 'Major Developments in Social Psychology During the Past Five Decades' in G. Lindzey and E. Aronson, (eds) *The Handbook of Social Psychology*, 3rd ed, vol. 1, New York: Random House, pp. 47-107.

Kahneman, D. and Tversky, A. (1979) 'Prospect Theory: an Analysis of Decision Making under Risk', *Econometrica*, vol. 47, pp. 263-291.

Kangethe, J. (1999) 'EIS Focus: Spatial Data Infrastructures in Africa', *EIS news*, October, pp. 15-20.

Kenny, D. A. (1985) 'Quantitative Methods for Social Psychology' in G. Lindzey and E. Aronson, (eds) *The Handbook of Social Psychology*, 3rd ed, vol. 1, New York: Random House, pp. 487-508.

Kenny, D. A., Kashy, D. A., and Bolger, N. (1998) 'Data Analysis in Social Psychology' D.T. Gilbert, S.T. Fiske, and G. Lindzey, (eds) *The Handbook of Social Psychology*, 4th ed, vol. 2, pp. 233-265.

Keren, G. (1996) 'Perspectives of Behavioral Decision Making: Some Critical Notes', *Organizational Behavior and Human Decision Processes*, vol. 65(3), pp. 169-178.

Kevany, M. J. (1995) 'A Proposed Structure for Observing Data Sharing' in H. J. Onsrud and G. Rushton, (eds) *Sharing Geographic Information*, New Brunswick, NJ: Center for Urban Policy Research, pp. 76-100.

King, J. L. (1995) 'Problems in Public Access Policy for GIS Databases: An Economic Perspective' in H. J. Onsrud and G. Rushton, (eds) *Sharing Geographic Information*, New Brunswick, NJ: Center for Urban Policy Research, pp. 255-276.

Kline, S. J. and Rosenberg, N. (1986) 'An Overview of Innovation' in R. Landau and N. Rosenberg, (eds) *The positive sum strategy*, Washington, D.C.: National Academy Press.

Klostermann, R. E. (1995) 'The Appropriateness of Geographic Information Systems for Regional Planning in the Developing World', *Computers, Environment, and Urban Systems*, vol. 19(1), pp. 1-13.

Knoke, D. and Janowiec-Kurle (1999) 'Make or Buy? The Externalization of Company Job Training' in S. B. Andrews and D. Knoke, (eds) *Networks in and around Organizations*, vol. 16, *Research In the Sociology of Organizations*, Stamford, Conn: Jai Press Inc., pp. 85-106.

Koh, W. T. H. (1992) 'Human Fallibility and Sequential Decision Making - Hierarchy versus Polyarchy', *Journal of Economic Behavior and Organization*, vol. 18, pp. 317-345.

Kondo, T. (1990) 'Some Notes on Rational Behavior; Normative Behavior; Moral Behavior; and Cooperation', *Journal of Conflict Resolution*, vol. 34(3), pp. 495-530.

Kuhl, J. (1985) 'From Cognition to Behavior: Perspectives for Future Research on Action Control' in J. Kuhl and J. Beckmann, (eds) *Action Control - From Cognition to Behavior*, Berlin: Springer-Verlag, pp. 267-275.

Kuhl, J. and Beckmann, J. (1985) 'Introduction and Overview' in J. Kuhl and J. Beckmann, (eds) *Action Control - From Cognition to Behavior*, Berlin: Springer-Verlag, pp. 1-8.

Laroche, H. (1995) 'From Decision to Action in Organizations: Decision-making as a Social Representation', *Organization Science*, vol. 6(1), pp. 62-75.

Lester, K. (1999) 'Using GIS to Establish a Corporate Database', presented at *Earth Data Information Systems Conference*, CSIR, Pretoria, 12-14 July.

Liebenberg, L. (1999) *The IDP Legislative Framework - A User Friendly Guide to Legislation dealing with Integrated Development Planning*, Cape Town: Foundation for Contemporary Research.

Lindblom, C. (1990) *Inquiry and Change - The Troubled Attempt to Understand and Shape Society*, New Haven and London; New York: Yale University Press and Russell Sage Foundation.

Liska, A. (1984) 'A Critical Examination of the Causal Structure of the Fishbein/Ajzen Attitude-Behavior Model', *Social Psychology Quarterly*, vol. 47(1), pp. 61-74.

Loewenthal, K. M. (1996) *An Introduction to Psychological Tests and Scales*, London: UCL Press Limited.

Londis, J. (1993) 'GIS Capabilities, Uses and Organisational Issues' in G. Castle, (ed.) *Profiting from Geographic Information Systems*, Fort Collins, Co: GIS World, Inc., pp. 43-53.

Longley, P. A., Goodchild, M. F., Maguire, D. J., and Rhind, D. W. (1999) 'Introduction' in P. A. Longley, M. F. Goodchild, D. J. Maguire, and D. W. Rhind, (eds) *Geographical Information Systems - Principles and Technical Issues*, vol. 1, New York: John Wiley & Sons, pp. 1-20.

MacDevette, D. R., Fincham, R. J., and Forsyth, G. G. (1999) 'The Rebuilding of a Country: the Role of GIS in South Africa' in P. A. Longley, M. F. Goodchild, D. J. Maguire, and D. W. Rhind, (eds) *Geographical Information Systems - Principles and Technical Issues*, vol. 2, New York: John Wiley & Sons, pp. 913-924.

Maguire, D. J. (1991) 'An Overview and Definition of GIS' in D. J. Maguire, M. F. Goodchild, and D. W. Rhind, (eds) *Geographical Information Systems - Principles and Applications*, vol. 1, Harlow: Longman Scientific & Technical, pp. 9-20.

Maguire, D. J., Goodchild, M. F., and Rhind, D. W. (eds) (1991) *Geographical Information Systems - Principles and Applications*, vol. 1 and 2, Harlow: Longman Scientific and Technical.

Mapping Science Committee (MSC) (1993) *Toward a Coordinated Spatial Data Infrastructure for the Nation*, Washington, DC: National Academy Press.

Mapping Science Committee (MSC) (1994) *Promoting the National Data Infrastructure Through Partnerships*, Board on Earth Sciences and Resources, Commision on Geosciences Environment and Resources, and National Research Council, Washington, DC.

Marble, D. E. and Peuquet, D. J. (1993) 'The Computer and Geography: Ten Years Later', *Professional Geographer*, vol. 45(4), pp. 446-448.

Markus, M. L. and Robey, D. (1988) 'Information Technology and Organizational Change: Causal Structure in Theory and Research', *Management Science*, vol. 34(5), pp. 583-598.

Martin, D. (1996) *Geographic Information Systems - Socioeconomic Applications*, 2nd ed, London: Routledge.

Martin, L. (1999) 'An IEC Case Study - The Technology behind SA's Democracy', *Network Times*, vol. 7(3-16), pp. 3-18.

Masser, I. (1999) 'All Shapes and Sizes: the First Generation of National Spatial Data Infrastructures', *International Journal of Geographical Information Science*, vol. 13(1), pp. 67-84.

Masser, I. and Blakemore, M. (eds) (1991) *Handling Geographic Information: Methodology and Potential Applications*, Harlow: Longman Scientific & Technical.

Masser, I. and Onsrud, H. J. (eds) (1993) *Diffusion and Use of Geographic Information Technologies*, Dordrecht: Kluwer Academic Publishers.

Medin, D. L. and Bazerman, M. H. (1999) 'Broadening Behavioral Decision Research: Multiple Levels of Cognitive Processing', *Psychonomic bulletin & review*, vol. 6(4), pp. 533-546.

Medyckyj-Scott, D. and Hearnshaw, H. M. (eds) (1993) *Human Factors in Geographic Information Systems*, London and Florida, Belhaven Press.

Mellers, B. A., Schwartz, A., and Cooke, D. J. (1998) 'Judgement and Decision Making', *Annual Review of Psychology*, vol. 49, pp. 447-477.

Metcalfe, J. S. (1981) 'Impulse and Diffusion in the Study of Technical Change', *Futures*, October, pp. 347-359.

Metcalfe, J. S. (1988) 'The Diffusion of Innovations: an Interpretative Survey' in G. Dosi, C. Freeman, R. Nelson, G. Silverberg, and L. Soete, (eds) *Technical Change and Economic Theory*, London and New York: Pinter Publishers, pp. 560-589.

Metselaar, E. E. (1997) *Assessing the Willingness to Change - Construction and validation of the DINAMO*, Amsterdam: University of Amsterdam.

Mizruchi, M. S. and Galaskiewicz, J. (1993) 'Networks of Interorganisational Relations', *Sociological Methods & Research*, vol. 22(1), pp. 46-70.

Mohr, L. B. (1982) *Explaining Organizational Behavior*, San Francisco: Jossey-Bass Publishers.

Monmonier, M. (1993) 'What a Friend We Have in GIS', *Professional Geographer*, vol. 45(4), pp. 448-450.

Montalvo Corral, C. (2002) *Environmental Policy and Technological Innovation: Why do firms adopt or reject new technologies?* Cheltenham, UK: Edward Elgar.

Moore, G. C. (1993) 'Implications from MIS Research for the Study of GIS Diffusion: Some Initial Evidence' in I. Masser and H. J. Onsrud, (eds) *Diffusion and Use of Geographic Information Technologies*, Dordrecht: Kluwer Academic Publishers, pp. 77-94.

Morgan, G. A. and Griego, O.V. (1998) *Easy Use and Interpretation of SPSS for Windows*, Mahwah, NJ: Lawrence Erlbaum Associates.

National Institute of Standards and Technology (NIST) (1992) *Federal Information Processing Standard*, Publication 173 (Spatial Data Transfer Standard), Washington, DC.: US Department of Commerce.

National Spatial Information Framework (NSIF) (1998a) 'The Way Forward: Developing a Framework Facilitating the Exchange and Utilisation of Geospatial Information', http://www.nsif.org.za/ (last accessed 30.7.1999).

National Spatial Information Framework (NSIF) (1998b) 'Interim Policy Guidelines', http://www.nsif.org.za/ (last accessed 21.6.2000).

Nebert, D. (ed.) (2000) *Developing Spatial Data Infrastructures: The SDI Cookbook,*Version 1.0, http://www.gsdi.org (last accessed 21.7.2000).

Nedovic-Budic, Z. (1998) 'The Impact of GIS technology', *Environment and Planning B: Planning and Design*, vol. 25, pp. 681-692.

Nedovic-Budic, Z. Pinto, J.K. (1999a) 'Interorganizational GIS: Issues and Prospects', *The Annals of Regional Science*, vol. 33, pp.183-195.

Nedovic-Budic, Z. Pinto, J.K. (1999b) 'Understanding Interorganizational GIS Activities: A Conceptual Framework', *Journal of the Urban and Regional Information Systems Association (URISA)*, vol.11(1), pp.53-64.

Nedovic-Budic, Z., Pinto, J.K. (2000) 'Information Sharing in an Interorganizational GIS Environment', *Environment and Planning B: Planning and Design*, vol. 27, pp. 455-474.

Nedovic-Budic. Z., Pinto, J.K. (2001) 'GIS Database Development and Exchanges: Interaction Mechanisms and Motivations', article submitted to *Journal of the Urban and Regional Information Systems Association (URISA)*, Version 09/24/01, www.urisa-online.org/journal

Nonaka, I. (1994) 'A Dynamic Theory of Organizational Knowledge Creation', *Organization Science*, vol. 5(1), pp. 14-37.

Nonaka, I., Takeuchi, H., and Umemoto, K. (1996) 'A Theory of Organizational Knowledge Creation', *International Journal of Technology Management, Special Issue on Unlearning and Learning for Technological Innovation*, vol. 11(7/8), pp. 833-845.

Nuttall, C. and Tunstall, D. (1996) 'AFRICAGIS'95 - Inventory of GIS Applications in Africa', United Nations Institute for Training and Research (UNITAR), World Resources Institute (WRI), Observatoire du Sahara et du Sahel (OSS), No. UNITAR/95/9, January.

Obermeyer, N. and Pinto, J. (1994) *Managing Geographic Information Systems*, New York and London: The Guildford Press.

Obermeyer, N. J. (1995) 'Reducing Inter-Organizational Conflict to Facilitate Sharing Geographic Information' in H. J. Onsrud and G. Rushton, (eds) *Sharing Geographic Information*, New Brunswick, NJ: Center for Urban Policy Research, pp. 138-148.

OGC (Open GIS Consortium) (1999) 'Open GIS Consortium, Inc.' , http://www.ogis.org/ (last accessed 21.6.2000).

Onrsud, H. and Rushton, G. (1996) *Institutions Sharing Geographic Information*, National Center for Geographic Information and Analysis, Closing Report, Research Initiative 9, March.

Onrsud, H. J., Johnson, J., Kirby, P., Moreno, R., and Ramlal, B. (1994) 'Land Information Systems in Developing Countries: Bibliography', National Center for Geographic Information and Analysis, Department of Surveying and Engineering, University of Maine, Orono, Maine, Report, No. 94-3.

Onsrud, H. J. (1995) 'The Role of Law in Impeding and Facilitating the Sharing of Geographic Information' in H. J. Onsrud and G. Rushton, (eds) *Sharing Geographic Information*, New Brunswick, NJ: Center for Urban Policy Research, pp. 292-306.

Onsrud, H. J. (1999) 'Survey of National Spatial Data Infrastructures - Compiled Responses by Question for Selected Countries', http://www.spatial.main.edu/~onsrud/GSID.htm (last accessed 21.6.2000).

Onsrud, H. J. and Pinto, J. K. (1991) 'Diffusion of Geographic Information Innovations', *International Journal of Geographical Information Systems*, vol. 5(4), pp. 447-467.

Onsrud, H. J. and Rushton, G. (1992) *NCGIA Research Initiative 9: Institutions Sharing Geographic Information*, National Center for Geographic Information and Analysis, Report of the Specialist Meeting, Technical Report 92-5, June.

Onsrud, H. J. and Rushton, G. (eds) (1995) *Sharing Geographic Information,* New Brunswick, NJ: Center for Urban Policy Research.

Openshaw, S. (1998) 'A New Human Geographic Research Agenda for GIS', *Environment and Planning*, vol. 30(3), pp. 383-384.

Openshaw, S., Charlton, M., and Carver, S. (1991) 'Error Propagation: a Monte Carlo Simulation' in I. Masser and M. Blakemore, (eds) *Handling Geographical Information: Methodology and Potential applications*, New York: Longman Scientific & Technical, pp. 78-101.

Opp, K.-D. (1999) 'Contending Conceptions of the Theory of Rational Action', *Journal of Theoretical Politics*, vol. 11(2), pp. 171-202.

Oppenheim, A. N. (1992) *Questionnaire Design, Interviewing and Attitude Measurement*, London and New York: Pinter Publishers.

Parker, D., Manstead, A. S. R., and Stradling, S. G. (1995) 'Extending the Theory of Planned Behaviour: The Role of Personal Norm', *British Journal of Social Psychology*, vol. 34, pp. 127-137.

Paul, B. K. (1993) 'A Case for Greater Interaction Between the Geographers of Developed and Developing Countries', *Professional Geographer*, vol. 45(4), pp. 461-465.

Payne, J. W. and Bettman, J. R. (1992) 'Behavioral Decision Research: A Constructive Processing Perspective', *Annual Review of Psychology*, vol. 43, pp. 87-131.

Petty, R. E. and Wegener, D. T. (1998) 'Attitude Change: Multiple Roles for Persuasion Variables' in D.T. Gilbert, S.T. Fiske, and G. Lindzey (eds) *The Handbook of Social Psychology*, 4th ed, vol. 1, pp. 323-390.

Peuquet, D. and Marble, D. (eds) (1990) *Introductory Readings in Geographic Information Systems,* London: Taylor & Francis.

Pfeffer, J. (1985) 'Organizations and Organization Theory' in G. Lindzey and E. Aronson, (eds) *The Handbook of Social Psychology*, 3rd ed, vol. 1, New York: Random House, pp. 379-440.

Pfeffer, J. (1997) *New Directions for Organization Theory - Problems and Prospects*, New York and Oxford: Oxford University Press.

Pfeffer, J. (1998) 'Understanding Organizations: Concepts and Controversies' *The Handbook of Social Psychology*, 4th ed, vol. 2, pp. 733-777.

Pfeffer, J. and Salancik, G. R. (1978) *The External Control of Organizations - A Resource Dependence Perspective*, New York: Harper & Row Publisher.

Pickles, J. (1993) 'Discourse on Method and the History of Discipline: Reflections on Dobson's 1983 Automated Geography', *Professional Geographer*, vol. 45(4), pp. 451-455.

Pickles, J. (ed.) (1995) *Ground Truth - The Social Implications of Geographic Information Systems*, New York: The Guildford Press.

Piernaar, M. and van Brakel, P. (1999) 'The Changing Face of Geographic Information on the Web: a Breakthrough in Spatial Data Sharing', *The Electronic Library*, vol. 17(6), pp. 365-371.

Pinto, J. K. and Onsrud, H. J. (1995) 'Sharing Geographic Information Across Organizational Boundaries: A Research Framework' in H. J. Onsrud and G. Rushton, (eds) *Sharing Geographic Information*, New Brunswick, NJ: Center for Urban Policy Research, pp. 44-64.

Plous, S. (1993) *The Psychology of Judgement and Decision Making*, New York: McGraw-Hill, Inc.

Polanyi, M. (1966) *The tacit dimension*, London: Routledge & Kegan Paul.

Porter, M. E. (1991) 'Towards a Dynamic Theory of Strategy', *Strategic Management Journal*, vol. 12, pp. 95-117.

Posey, A. S. (1993) 'Automated Geography and the Next Generation', *Professional Geographer*, vol. 45(4), pp. 455-456.

Rhind, D. (ed.) (1997) *Framework for the World*, Cambridge: GeoInformation International.

Rhind, D. (1998) 'Public/private sector relationships in the Creation, Management and Exploitation of Geospatial Data' in Conference Proceedings, *XXI International Congress Developing the Profession in a Developing World - Commision 3: Land Information Systems*, vol. 3, Brighton: The International Federation of Surveyors, pp. 201-215.

Rhind, D. W. (1992) 'Data Access, Charging and Copyright and their Implications for GIS', *International Journal of Geographical Information Systems*, vol. 6, pp. 13-30.

Rhind, D. W. (1999) 'National and International Geospatial Data Policies' in P. Longley, M. Goodchild, D. Maguire, and D. W. Rhind, (eds) *Geographical Information Systems - Management Issues and Applications*, vol. 2, New York: John Wiley & Sons, pp. 767-787.

Rodeghier, M. (1996) *Surveys with Confidence - A Practical Guide to Survey Research using SPSS*, Chicago: SPSS.

Rogers, E. M. (1968) *Diffusion of Innovations*, New York: The Free Press.

Rogers, E. M. (1993) 'The Diffusion of Innovations Model: Keynote Address' in I. Masser and H. J. Onsrud, (eds) *Diffusion and Use of Geographic Information* Technologies, Dordrecht: Kluwer Academic Publishers, pp. 9-24.

Rogers, E. M. (1995) *Diffusion of Innovations*, 4th ed, New York: The Free Press.

Rosenberg, N. (1976) 'Factors Affecting the Diffusion of Technology' in N. Rosenberg, (ed.) *Perspectives on Technology*, Cambridge: Cambridge University Press, pp. 189-210.

Sabini, J. (1995) *Social Psychology*, 2nd ed, New York: W.W. Norton & Company.

Sanbonmatsu, D. M., Posavac, S. S., and Stasney, R. (1997) 'The Subjective Beliefs Underlying Probability Overestimation', *Journal of Experimental Social Psychology*, vol. 33, pp. 276-295.

Sarkar, J. (1998) 'Technological Diffusion: Alternative Theories and Historical Evidence', *Journal of Economic Surveys*, vol. 12(2), pp. 131-176.

Sarver, V. T., Jr. (1983) 'Ajzen and Fishbein's "Theory of Reasoned Action": A Critical Assessment', *Journal for the Theory of Social Behavior*, vol. 13, pp. 155-163.

Savage, D., Whelan, P., Liebenberg, L., and Malan, L. (1999) 'Review of Integrated Development Planning in the Western Cape', Foundation for Contemporary Research, Cape Town, November 1998.

Sayer, A. (1992) *Method in Social Science - A Realist Approach*, 2nd ed, London and New York: Routledge.

Schoemaker, P. J. H. (1993) 'Strategic Decisions in Organizations: Rational and Behavioural Views', *Journal of Management Studies*, vol. 30(January), pp. 107-129.

Schuman, H. and Kalton, G. (1985) 'Survey Methods' in G. Lindzey and E. Aronson, (eds) *The Handbook of Social Psychology*, 3rd ed, vol. 1, New York: Random House.

Schuman, H. and Presser, S. (1981) *Questions and Answers in Attitude Surveys - Experiments on Question Form, Wording, and Context*, New York: Academic Press.

Schwabe, C., O'Leary, B., Sukai, S.B. (1998) 'Putting all the Facts on the Map', *In Focus Forum*, vol. 5(3), pp. 4-7.

Shepherd, J. (1991) 'Planning Settlements and Infrastructure' in I. Masser and M. Blakemore, (eds) *Handling Geographical Information: Methodology and Potential Applications*, New York: Longman Scientific & Technical, pp. 181-220.

Sheppard, E. (1993) 'Automated Geography: What Kind of Geography for What Kind of Society?', *Professional Geographer*, vol. 45(4), pp. 457-460.

Shiffer, M. J. (1999) 'Managing a Whole Economy: the Contribution of GIS' in P. A. Longley, M. F. Goodchild, D. J. Maguire, and D. W. Rhind, (eds) *Geographical Information Systems - Principles and Technical Issues*, vol. 2, New York: John Wiley & Sons, pp. 723-732.

Smith, N. S. and Rhind, D. W. (1999) 'Characteristics and Sources of Framework Data' in P. Longley, M. Goodchild, D. Maguire, and D. W. Rhind, (eds) *Geographical Information Systems - Management Issues and Applications*, vol. 2, New York: John Wiley & Sons, pp. 655-666.

Smith, R. (1998) 'Southern African Metadata Consortium (SAM)', in D. Clarke, E. Gavin, W. Honu, T. Krieg, M. Muller, H.J. Smith, T. Smith, and S. Vorster (eds) *National Spatial Information Framework Workshop*, proceedings, Sinodale Sentrum, Pretoria, 11 February, pp. 24-26.

Sommer, B. and Sommer, R. (1991) *A Practical Guide to Behavioral Research - Tools and Techniques*, New York and Oxford: Oxford University Press.

Sperling, J. (1995) 'Development and Maintenance of the TIGER Database: Experiences in Spatial Data Sharing at the U.S. Bureau of the Census' in H. J. Onsrud and G. Rushton, (eds) *Sharing Geographic Information*, New Brunswick, NJ: Center for Urban Policy Research, pp. 377-396.

Stahlberg, D. and Frey, D. (1996) 'Attitudes: Structure, Measurement and Functions' in M. Hewstone, W. Stroebe, and G. Stephenson, (eds) *Introduction to Social Psychology*, 2nd ed, Oxford: Blackwell Publishers.

Star, J. and Estes, J. (1990) *Geographic Information Systems - An Introduction*, Englewood Cliffs, NJ: Prentice Hall.

Staw, B. M. (1991) 'Dressing Up Like an Organization: When Psychological Theories Can Explain Organizational Action', *Journal of Management*, vol. 17(4), pp. 805-819.

Stevenson, M. K., Busemeyer, J. R., and Naylor, J. C. (1990) 'Judgement and Decision-making Theory' in M. D. Dunette and L. M. Hough, (eds) *Handbook of industrial and organizational psychology*, 2nd ed, vol. 1, Palo Alto, CA: Consulting Psychologists Press, pp. 283-367.

Sudman, S. and Bradburn, N. M. (1982) *Asking Questions*, San Francisco: Jossey-Bass Publishers.

Sutton, S. (1998) 'Predicting and Explaining Intentions and Behavior: How Well Are We Doing?', *Journal of Applied Social Psychology*, vol. 28(15), pp. 1317-1338.

Svenson, O. (1996) 'Decision Making and the Search for Fundamental Psychological Regularities: What Can Be Learned from a Process Perspective?', *Organizational Behavior and Human Decision Processes*, vol. 65(3), pp. 252-267.

Swan, J. A. (1995) 'Exploring Knowledg and Cognitions in Decisions About Technological Innovation: Mapping Managerial Cognitions', *Human Relations*, vol. 48(11), pp. 1241-1270.

Swarts, M. (1998) 'Nutshell', *In Focus Forum*, vol. 5(3), pp. 2.

Taupier, R. P. (1995) 'Comments on the Economics of Geographic Information and Data Access in the Commonwealth of Massachusetts' in H. J. Onsrud and G. Rushton, (eds) *Sharing Geographic Information*, New Brunswick, NJ: Center for Urban Policy Research, pp. 277-291.

Taylor, D. R. F. (1991) 'GIS and Developing Nations' in D. Maguire, M. Goodchild, and D. Rhind, (eds) *Geographical Information Systems - Principles and Applications*, vol. 2, Harlow: Longman Scientific and Technical, pp. 71-84.

Taylor, D. W. (1965) 'Decision Making and Problem Solving' in J. March, (ed.) *Handbook of Organizations*, Chicago: Rand McNally, pp. 48-86.

Taylor, P. J. and Johnston, R. J. (1995) 'Geographic Information Systems and Geography' in J. Pickles, (ed.) *Ground Truth - The Social Implications of Geographic Information Systems*, New York and London: The Guildford Press, pp. 51-67.

Taylor, S. and Todd, P. A. (1995) 'Understanding Information Technology Usage: A Test of Competing Models', *Information Systems Research*, vol. 6(2), pp. 144-175.

Terry, D. J. and O'Leary, J. E. (1995) 'The Theory of Planned Behaviour: The Effects of Perceived Behavioural Control and Self-efficacy', *British Journal of Social Psychology*, vol. 34, pp. 199-220.

Tosta, N. (1995) 'The Evolution of Geographic Information Systems and Spatial Data-Sharing Activities In California State Government' in H. J. Onsrud and G. Rushton, (eds) *Sharing Geographic Information*, New Brunswick, NJ: Center for Urban Policy Research, pp. 193-206.

Trafimow, D. and Fishbein, M. (1995) 'Do People Really Distinguish Between Behavioural and Normative Beliefs?', *British Journal of Social Psychology*, vol. 34, pp. 257-266.

Unwin, D. (1981) *Introductory Spatial Analysis*, London and New York: Methuen.

van Helden, P. (1999) 'GIS for Development: An Analysis of GIS Diffusion in Government Organisations in South Africa', presented at *Earth Data Information Systems Conference*, CSIR, Pretoria, 12-14 July.

van Helden, P. (2000) 'GIS Adoption in South Africa Government', *GEO Informatics*, vol. 3(January/February), pp. 55.

Ventura, S. J. (1995) 'Overarching Bodies for Coordinating Geographic Data Sharing at Three Level of Government' in H. J. Onsrud and G. Rushton, (eds) *Sharing Geographic Information*, New Brunswick, NJ: Center for Urban Policy Research, pp. 172-192.

Veregin, H. (1995) 'Computer Innovation and Adoption in Geography' in J. Pickles, (ed.) *Ground Truth - The Social Implications of Geographic Information Systems*, New York and London: The Guildford Press, pp. 88-112.

Webster (1960) *Webster's New Collegiate Dictionary*, Springfield, MA: G. & C. Merriam Co. Publishers.

Webster, C. J. (1994) 'GIS and the scientific inputs to planning - Part 2: prediction and prescription', *Environment and Planning B: Planning and Design*, vol. 21, pp. 145-157.

Wilson, J. D. (1999) 'Stellar performers - Top companies vie for position in the GIS universe', *GEOEurope* (March), pp. 36-41.

World Bank (1999) *World Development Report - Knowledge for Development*, Oxford: Oxford University Press.

Yates, J. F. (1990) *Judgement and Decision Making*, Englewood Cliffs, NJ: Prentice Hall.

Index